T0181948

Pillars of Transcendental Number Theory

This picture is the great corridor at Rameshwaram Temple, Tamilnadu. The picture is taken from the following url address: https://commons.wikimedia.org/wiki/File:Grand_corridor, rameshwaram_temple,tamilnadu_-_panoramio.jpg. Our thanks to Mr. Rajaraman Sundaram who has taken the photograph and made available for public in the above website

Saradha Natarajan · Ravindranathan Thangadurai

Pillars of Transcendental Number Theory

Saradha Natarajan
DAE Centre for Excellence
in Basic Sciences
University of Mumbai
Mumbai, Maharashtra, India

Ravindranathan Thangadurai
Department of Mathematics
Harish-Chandra Research Institute
Prayagraj, Uttar Pradesh, India

ISBN 978-981-15-4157-5   ISBN 978-981-15-4155-1   (eBook)
https://doi.org/10.1007/978-981-15-4155-1

This Springer imprint is published by the registered company Springer Nature Singapore Pte Ltd.
The registered company address is: 152 Beach Road, #21-01/04 Gateway East, Singapore 189721, Singapore

I seem to have been only like a boy playing on the seashore, and diverting myself in now and then finding a smoother pebble or a prettier shell than ordinary, whilst the great ocean of truth lay all undiscovered before me

—Isaac Newton

Dedicated

To

Our families

Preface

Since the proof of Hermite in 1873 on the transcendence of the classical constant e, the theory of transcendence has evolved as a well-developed subject due to the contributions of many mathematicians like Lindemann, Thue, Siegel, Gelfond, Schneider, Roth and others. Baker's work on linear forms in logarithms from 1965 to 1970 gave an impetus to the subject. The years that followed saw a resurgence of the subject. Several improvements were made in the lower bound estimates for linear forms in logarithms by Baker himself, Waldschmidt, Shorey and many others. These improved bounds have a wide number of *effective* applications in Diophantine equations, class number problems, powers in recurrence sequences, etc. Numerous papers have been written and are still being written using the theory of linear forms. The theory is a gold mine for researchers. There are several mathematicians all around the world who have made and are still making important contributions to this wonderful subject. We have mentioned only very few names and fewer results to keep the book as simple as possible.

There are a handful of books written on the theory of linear forms and its applications beginning with the book of Baker [1]. Some of the other books are by Waldschmidt [2], Shorey & Tijdeman [3], Baker & Wüstholz [4], Nesterenko [5] and Ram Murty & Rath [6].

With the ever-growing applications, it becomes important for any student who wants to work in this area, to know the proofs of Baker's original results. For this purpose, the only available sources are either Baker's book or his original papers or some online notes.

One of our primary aims of writing this book is to present Baker's original results in a way, suitable for students of postgraduate or first-year Ph.D. level. We intend to keep the exposition simple and easily accessible.

This book will begin with some classical results like the transcendence of e, π and Hermite–Lindemann–Weierstrass theorem. Our proof of the Gelfond–Schneider theorem is based on Siegel's method. A new feature here is that we will be showing some well-known results of Ramachandra, which are not widely known. Gelfond–Schneider theorem and many other interesting results will be derived from his results.

An important area, which got an impetus due to an ingenious method initiated by Thue in 1909, was *Diophantine Approximation*. This was developed by Siegel, Dyson, Gelfond and finally Roth, in 1955, obtained the best possible result for the approximation of an algebraic number by a rational. This result is now known as Thue–Siegel–Roth theorem. This theorem is an important pillar in this subject. The proof of Roth's theorem can be found in Schmidt [7]. There are a few other books which outline the proof of Roth's theorem. We will give proofs of theorems of Thue, improvement by Siegel and the theorem of Roth in this book. We will follow the original paper of Roth [8] (and [9]) for the proof of his theorem.

When students see all the three proofs in one place, they can understand how the ideas developed. Our aim is not only to give the proofs of the theorems, but to illustrate the *ineffectiveness* of these theorems with the *effectiveness* of Baker's result in solving Diophantine equations. Thue's equations have attracted a lot of attention lately. There are computational programs developed, by which, nowadays it is a routine matter to solve a Thue's equation. For any student who wants to work in this and allied areas, this is a prerequisite.

Another important pillar in this subject is Schmidt's subspace theorem. This is a multidimensional analogue of Roth's theorem. Of late, many interesting applications of subspace theorem are being discovered. Although we will not be able to give the proof of this theorem, we will illustrate the theorem with some applications including a recent one.

Each chapter will conclude with a few problems in the form of Exercise and some interesting information as Notes. We do not intend to be exhaustive in both Exercise and Notes. These are meant to infuse and instil curiosity for the reader.

Our aim is to bring important theorems of transcendence theory under one roof so that the subject can be taught as a well-knit graduate course. The style will be classical, simple and friendly to students. Postgraduate and Ph.D. students of pure mathematics can be benefited by a course based on this book. It will be a good collection in any pure mathematician's library and can use this for teaching a course in transcendence. As the book will be self-contained, the need to refer to other books will be minimal. A basic course in algebraic number theory [10], real and complex analysis [11, 12] will be required for any reader of this book.

The first author would like to thank the Indian National Science Academy for awarding her Senior Scientist Fellowship and DAE Centre for Excellence in Basic Sciences, University of Mumbai, for providing facilities. The second author is thankful to Harish-Chandra Research Institute for the excellent environment which helped in writing the book, and also he is thankful to Dr. Veekesh Kumar and Ms. Bidisha Roy for going through some chapters of this book in the first draft and for suggesting some of the exercises.

Mumbai, India   Saradha Natarajan
Prayagraj, India   Ravindranathan Thangadurai

References

1. A. Baker, *Transcendental Number Theory* (Cambridge Tracts, 1975) (Preface, Chapter 7 and Chapter 8).
2. M. Waldschmidt, *Nombres Transcendants* (Springer-Verlag, 1974) (Preface and Chapter 4).
3. T.N. Shorey, R. Tijdeman, *Exponential Diophantine Equations* (Cambridge Tracts, 1986 and re-printed in 2008) (Preface and Chapter 7).
4. A. Baker, G. Wüstholz, *Logarithmic Forms and Diophantine Geometry* (Cambridge Tracts, 2007) (Preface and Chapter 7).
5. Y.V. Nesterenko, *Algebraic Independence*, vol. 14 (Tata Institute of Fundamental Research Publications, 2008), 157pp (Preface, Chapter 2 and Chapter 3).
6. M. Ram Murty, P. Rath, *Transcendental Numbers* (Springer, 2014), 217pp (Preface, Chapter 2, Chapter 4 and Chapter 7).
7. W.M. Schmidt, *Diophantine Aproximation*, vol. 785 (Springer-Verlag LNM, Berlin-Heidelberg, 1980) (Preface and Chapter 9).
8. K.F. Roth, *Rational Approximations to Algebraic Numbers*. Mathematika **2**, 1–20 (1955); *Corrigendum* **2** (1955), p. 168. (Preface and Chapter 6).
9. W.J. LeVeque, *Topics in Number Theory, Vol I and II* (Dover Publication Inc, New York, 1984) (Preface and Chapter 6).
10. S. Lang, *Algebraic Number Theory*, vol. 110, 2nd edn. Graduate Texts in Mathematics (Springer-Verlag, New York, 1994) (Preface and Chapter 1).
11. T.M. Apostol, *Mathematical Analysis: A Modern Approach to Advanced Calculus* (Addison-Wesley Publishing Company, Inc., Reading, Mass., 1957) (Preface).
12. W. Rudin, *Real and Complex Analysis*, 3rd edn (McGraw-Hill Book Co., New York, 1987) (Preface).

Contents

About the Authors

Saradha Natarajan is an INSA Senior Scientist at the DAE Center for Excellence in Basic Sciences at the University of Mumbai, India, and elected fellow of the Indian National Science Academy (INSA). Earlier, she was Professor of Mathematics at the Tata Institute of Fundamental Research, Mumbai, India, until 2016. She earned her Ph.D. in 1983 under the guidance of Prof. T. S. Bhanumurthy from the Ramanujan Institute for Advanced Study in Mathematics, University of Madras, Chennai. She was a postdoctoral fellow at Concordia University, Canada; Macquarie University, Australia; and National Board of Higher Mathematics (NBHM), India.

Her area of specialization is number theory, in general, and transcendental number theory and Diophantine equations, in particular. She has published several papers in international journals of repute and has collaborated with many mathematicians both in India and abroad. Several students have completed their Ph.D. under her supervision. She has travelled extensively and given invited talks and lectures at national and international seminars and conferences. Professor Natarajan has made substantial contributions to the conjectures of Erdos on perfect powers in arithmetic progressions, where combinatorial and computational methods, linear forms in logarithms and modular method are combined. She also has made significant contributions to Thue equations and Diophantine approximations, especially towards conjectures of Bombieri, Mueller and Schmidt on number of solutions of Thue inequalities for forms in terms of number of non-zero coefficients of the form. In the area of transcendence, she has obtained best possible simultaneous approximation measures for values of exponential function and Weierstrass elliptic function. Further, significant lower bounds were shown for the Ramanujan tau-function for almost all primes p.

Ravindranathan Thangadurai is Professor at Harish-Chandra Research Institute, Prayagraj, India. He earned his Ph.D. in Combinatorial Number Theory in 1999 from the Mehta Research Institute for Mathematics and Theoretical Physics, Allahabad (now Harish-Chandra Research Institute, Prayagraj) under the supervision of Prof. S. D. Adhikari. He spent two years as a postdoc at the Institute of Mathematical Sciences, Chennai, India, and two years at Indian Statistical Institute, Kolkata, India.

His areas of research include analytic, combinatorial and transcendental number theory, specifically, major contributions in the area of zerosum problems in finite abelian groups, distribution of residues modulo p, Liouville numbers and Schanuel's conjecture in transcendental number theory. He has collaborated with reputed mathematicians and his research articles have been published journals of repute. He has computed the exact values of Olson's constant and Alon–Dubiner constant for subsets for the group. He proved a conjecture of Schmid and Zhuang for large class of finite abelian p-groups and the current best known upper bound for Davenport's constant for a general finite abelian group. He has also made a major contribution to the theory of distribution of particular type of elements (specially, quadratic non-residues but not a primitive root) of residues modulo p. He has proved a strong form of Schanuel's conjecture in transcendental number theory for many n-tuples.

Symbols

$\mathbb{N}$	Positive rational integers
$\mathbb{Z}$	Ring of rational integers
$\mathbb{Q}$	Field of rational numbers
$\mathbb{R}$	Field of reals
$\mathbb{C}$	Field of complex numbers
$\mathbb{A}$	Field of algebraic numbers
$\mathcal{K}$	Number field
$\mathcal{O}_\mathcal{K}$	Ring of integers of $\mathcal{K}$
$\mathcal{O}_\mathcal{K}^*$	Group of units of $\mathcal{O}_\mathcal{K}$
$\mathcal{N}(\alpha)$	Norm of α over $\mathbb{Q}$ with respect to $\mathbb{Q}(\alpha)$
$\mathcal{N}_{\mathcal{K}/\mathbb{Q}}(\alpha)$	norm of α over $\mathbb{Q}$ with respect to $\mathcal{K}$
$\mathbf{F}[X]$	Ring of polynomials in X with coefficients in $\mathbf{F}$
$\mathbf{F}[X_1,\ldots,X_n]$	Ring of polynomials in $X_1,\ldots,X_n$ with coefficients in $\mathbf{F}$
$\mathcal{K}[[z]]$	Ring of power series with coefficients in $\mathcal{K}$
$[x]$	Integral part of real number x
$\{x\}$	Fractional part of real number x
$||x||$	Distance of real number x to the nearest integer
$\Re(z)$	Real part of z
$\Im(z)$	Imaginary part of z
$H(P)$	Height of a polynomial P
$h(\alpha)$	Height of an algebraic number α
$\lceil\alpha\rceil$	House of an algebraic number α
$d(\alpha)$	Denominator of an algebraic number α
$s(\alpha)$	Size of an algebraic number α
$h^\circ(\alpha)$	Absolute logarithmic height of an algebraic number α

Chapter 1
Preliminaries

The shell must break before the bird can fly

—Tennyson

In this chapter, we collect some basic results which will be used in the ensuing chapters.

While looking for the transcendence nature of values of some classical functions, it is apriori imperative to check if the given functions are transcendental functions or algebraically independent functions. In Sect. 1.1, we show the algebraic independence of the pairs of functions (z, e^{az}), (e^z, e^{az}) and $(\wp(z), \wp^*(z))$ where $0 \neq a \in \mathbb{C}$, $\wp$ and $\wp^*$ are Weierstrass elliptic functions.

Section 1.2 deals with the Gauss's lemma on primitive polynomials and factorisation of multi-variable polynomials.

In Sect. 1.3, we collect some basic properties of algebraic numbers and state Dirichlet's Unit theorem.

We give a criterion for linear independence of functions of single and several variables in terms of Wronskians in Sect. 1.4. In the multi-variable case, the notion of generalised Wronskian was introduced by Roth while proving his famous result on the approximation of an algebraic number by rationals. Their properties are explained in Lemmas 1.4.2 and 1.4.3. These are of independent interest also.

1.1 Algebraic Independence of Functions

Let $f_1, f_2, \ldots, f_\ell$ be complex functions defined on a field $\mathcal{K} \subset \mathbb{C}$. We say these functions are *algebraically independent over* $\mathcal{K}$ (respectively, *linearly independent over* $\mathcal{K}$), if for every non-zero polynomial $P(x_1, \ldots, x_\ell) \in \mathcal{K}[x_1, \ldots, x_\ell]$ (respectively, non-zero linear polynomial), there exists $z_0 \in \mathcal{K}$ such that the complex number

© Springer Nature Singapore Pte Ltd. 2020
S. Natarajan and R. Thangadurai, *Pillars of Transcendental Number Theory*,
https://doi.org/10.1007/978-981-15-4155-1_1

$P(f_1(z_0), \ldots, f_\ell(z_0)) \neq 0$. When $\ell = 1$, we say the function f_1 is *transcendental over* $\mathcal{K}$.

Let $\rho > 0$ be a real number. We say that an entire function f is of *order* ρ if there exists an absolute constant $C > 0$ such that

$$|f|_R = \max_{|z|=R} |f(z)| \leq C^{R^\rho} \text{ for } R \to \infty.$$

For example, any polynomial $P(z) \in \mathbb{C}[z]$ is of order 0 while e^z has order 1. This notion is extended to meromorphic functions as follows. A meromorphic function is said to be of *order* ρ if it is the quotient of two entire functions of order $\leq \rho$. For example, the Weierstrass elliptic function $\wp(z)$ is known to be the quotient of entire functions of order 2. Hence $\wp(z)$ is of order 2.

As a consequence of well-known Jensen's formula, we get that the number of zeros of an entire function of order $\leq \rho$, inside a circle of radius R is at most $O(R^\rho)$ as $R \to \infty$. We will use these facts to show the algebraic independence of certain classical functions below. It is well known that e^z is a transcendental function. We show the following result.

Lemma 1.1.1 *For any non-zero $a \in \mathbb{C}$, the functions z and e^{az} are algebraically independent over $\mathbb{C}$.*

Proof Suppose the lemma is false. Then there exists a non-zero polynomial $P(x_1, x_2)$ such that $P(e^{az}, z) = 0$ for all $z \in \mathbb{C}$. We shall take the polynomial P to be of least degree in x_1. Let the degree of P in x_1 be ν. Thus there exist non-zero polynomials $f_0(z), \ldots, f_\nu(z) \in \mathbb{C}[z]$, not all constants such that $f_0(z) \not\equiv 0$ and

$$f_0(z)e^{\nu az} + f_1(z)e^{(\nu-1)az} + \cdots + f_\nu(z) = 0.$$

Divide out by $f_0(z)$ to get an equation of the form

$$Q(z) = e^{\nu az} + g_1(z)e^{(\nu-1)az} + \cdots + g_\nu(z) = 0$$

with $g_i(z) = f_i(z)/f_0(z)$ for $1 \leq i \leq \nu$. Note that $Q(z)$ is uniquely determined since ν is the least degree in x_1. We know that e^{az} is invariant under $z \to z + 2n\pi i/a, n \in \mathbb{Z}$. Thus each $g_i(z), 1 \leq i \leq \nu$ must be invariant under these transformations. That is, $g_i(z) = g_i(z + 2n\pi i/a)$. That means each $g_i(z)$ has either infinitely many zeros or poles. Since g_i is a rational function, we conclude each $g_i(z)$ is a constant which is a contradiction.   □

By similar argument, we can also show the following lemma.

Lemma 1.1.2 *The functions e^z and e^{az}, a irrational are algebraically independent over $\mathbb{C}$.*

Proof Suppose the lemma is false. Arguing as in Lemma 1.1.1, there exists a unique relation

$$Q(z) = e^{\nu az} + g_1(e^z)e^{(\nu-1)az} + \cdots + g_\nu(e^z) = 0$$

with $g_i(e^z) = f_i(e^z)/f_0(e^z)$ for $1 \le i \le \nu$. Again, each $g_i(e^z)$ must be invariant under the transformation $z \to z + 2n\pi i/a, n \in \mathbb{Z}$. Since a is irrational, each $g_i(e^z)$ has two independent periods $w_1 := 2\pi i$ and $w_2 := 2\pi i/a$ and so $\ell_1 w_1 + \ell_2 w_2$ is a period for any integers ℓ_1 and ℓ_2. Hence, the number of zeros or poles inside a circle of radius R of any $g_i(e^z)$ is bounded below by $O(R^2)$. This is a contradiction as the order of $g_i(e^z)$ is 1.   □

Another important function which has been widely studied is the Weierstrass elliptic function $\wp(z)$. This is a doubly periodic meromorphic function which is a quotient of entire functions of order 2. In fact, these entire functions are known as σ functions. For various properties of this function which will be used in this book, we refer to [1].

Lemma 1.1.3 *Let $\wp(z)$ and $\wp^*(z)$ be two elliptic functions with periods (ω_1, ω_2) and (ω_1^*, ω_2^*). Then $\wp$ and $\wp^*$ are algebraically dependent if and only if their periods are commensurable i.e there exists a 2×2 rational matrix M such that*

$$\begin{pmatrix} \omega_1 \\ \omega_2 \end{pmatrix} = M \begin{pmatrix} \omega_1^* \\ \omega_2^* \end{pmatrix}. \tag{1.1}$$

Proof Suppose there exists $M = \begin{pmatrix} a & b \\ c & d \end{pmatrix}$ satisfying (1.1) with $a, b, c, d \in \mathbb{Q}$. Then

$$\omega_1 = a\omega_1^* + b\omega_2^*; \quad \omega_2 = c\omega_1^* + d\omega_2^*.$$

Hence there exists an integer m such that

$$m\omega_1 = a'\omega_1^* + b'\omega_2^*; \quad m\omega_2 = c'\omega_1^* + d'\omega_2^*$$

with $a', b', c', d' \in \mathbb{Z}$. (One may take m to be the least common multiple of the denominators of a, b, c, d). Thus $\wp^*(mz)$ has fundamental periods ω_1 and ω_2. Hence $\wp^*(mz)$ is a rational function of $\wp(z)$. A priori, $\wp^*(mz)$ is a rational function of $\wp^*(z)$ which therefore implies that $\wp(z)$ and $\wp^*(z)$ are algebraically dependent.

Now we prove the converse. Suppose $\wp(z)$ and $\wp^*(z)$ are algebraically dependent. Arguing as in Lemma 1.1.1 there exists a unique relation

$$Q(z) = \wp(z)^\nu + g_1(\wp^*(z))\wp(z)^{(\nu-1)} + \cdots + g_\nu(\wp^*(z)) = 0$$

with $g_i(\wp^*(z)) = f_i(\wp^*(z))/f_0(\wp^*(z))$ for $1 \le i \le \nu$. Then each $g_i(\wp^*(z))$ is invariant under $z \to z + m\omega_1, m \in \{0, 1, 2, \ldots\}$ and under $z \to z + n\omega_2, n \in \{0, 1, 2, \ldots\}$. This means $\wp^*(z + m\omega_1)$ and $\wp^*(z + n\omega_2)$ are not all distinct. Hence there exist integers m_0 and n_0 such that $m_0\omega_1$ and $n_0\omega_2$ are periods of $\wp^*(z)$. Thus for some integers m_1, m_2, n_1, n_2 we have $m_0\omega_1 = m_1\omega_1^* + m_2\omega_2^*$ and $n_0\omega_1 = n_1\omega_1^* + n_2\omega_2^*$ which proves (1.1).   □

1.2 Gauss's Lemma

Let $P(z) \in \mathbb{Z}[z]$ be a polynomial. Let $C(P)$ denote the greatest common divisor of the coefficients of P, and it is called the *content* of P. We say $P(z)$ is a *primitive polynomial* if $C(P) = 1$. We show

Lemma 1.2.1 *The product of two primitive polynomials is a primitive polynomial. Thus for any two polynomials P, $Q \in \mathbb{Z}[z]$ we have $C(PQ) = C(P)C(Q)$.*

Proof Let $P(z)$ and $Q(z)$ be two primitive polynomials with coefficients in $\mathbb{Z}$. Suppose their product PQ is not primitive. Then there exists a prime p such that

$$P(z)Q(z) \text{ is identically zero in } \mathbb{Z}/p\mathbb{Z}[z],$$

which we write as $P(z)Q(z) \equiv 0 \pmod{p}$. Let $P_1(z)$ and $Q_1(z)$ be the polynomials $P(z)$ and $Q(z)$ reduced $\pmod{p}$., i.e. $P_1(z) \equiv P(z) \pmod{p}$ and $Q_1(z) \equiv Q(z)$ $\pmod{p}$. Hence

$$P_1(z)Q_1(z) \equiv 0 \pmod{p}. \tag{1.2}$$

Since $P(z)$ and $Q(z)$ are both primitive, $P_1(z)$ and $Q_1(z)$ are not identically 0. Let p_1 and q_1 be the leading coefficients of P_1 and Q_1, respectively. Then $p_1 \not\equiv 0$ $\pmod{p}$, $q_1 \not\equiv 0 \pmod{p}$ and hence $p_1 q_1 \not\equiv 0 \pmod{p}$. This contradicts (1.2). Thus PQ is primitive.

Any polynomial $P(z) \in \mathbb{Z}[z]$ can be written as $P(z) = C(P)P'(z)$ with P' primitive. If $R = PQ$, then $C(R)R' = C(P)C(Q)P'Q'$. Since $P'Q'$ is primitive, it follows that $C(R) = C(P)C(Q)$. □

It is possible to generalise the above result to any number of variables as follows.

Lemma 1.2.2 *Let P, $Q \in \mathbb{Z}[z_1, \ldots, z_m]$. Then*

$$C(PQ) = C(P)C(Q).$$

For a proof, we refer to Cassels [2]. It follows from the above lemmas that if P can be factored over $\mathbb{Q}$, it can also be factored over $\mathbb{Z}$.

Lemma 1.2.3 *Let $1 \le r < m$. Suppose that $F(z_1, \ldots, z_m) \in \mathbb{Z}[z_1, \ldots, z_m]$, $G(z_1, \ldots, z_r) \in \mathbb{Q}[z_1, \ldots, z_r]$ and $H(z_{r+1}, \ldots, z_m) \in \mathbb{Q}[z_{r+1}, \ldots, z_m]$ such that*

$$F(z_1, \ldots, z_m) = G(z_1, \ldots, z_r)H(z_{r+1}, \ldots, z_m).$$

Let γ be a coefficient in F. Then there is a factorisation $\gamma = \alpha\beta$ in $\mathbb{Q}$ such that $\alpha G \in \mathbb{Z}[z_1, \ldots, z_r]$ and $\beta H \in \mathbb{Z}[z_{r+1}, \ldots, z_m]$.

Proof Let the coefficients of G be $\alpha_1, \ldots, \alpha_s$ and that of H be $\beta_1, \ldots, \beta_t$ in some order. Since the variables in G and H are disjoint, the coefficients of F are $\alpha_i \beta_j$ and they are all in $\mathbb{Z}$. In particular, $\alpha_1 \beta_j \in \mathbb{Z}$ for $1 \le j \le t$ and $\beta_1 \alpha_j \in \mathbb{Z}$ for $1 \le j \le s$.

But these are coefficients of $\alpha_1 H$ and $\beta_1 G$, respectively, and $\alpha_1\beta_1$ is some coefficient of F. By similar argument, every coefficient of F can be written in the required form. □

1.3 Properties of Algebraic Numbers

A complex number $\alpha \in \mathbb{C}$ is said to be *an algebraic number* if there exists a non-zero polynomial $P(x) \in \mathbb{Z}[x]$ such that $P(\alpha) = 0$. If $P(x)$ is monic, then α is said to be *an algebraic integer*. A complex number $\alpha \in \mathbb{C}$ is said to be *a transcendental number* if α is not an algebraic number. The set of all algebraic numbers is denoted by $\mathbb{A}$.

For a given $\alpha \in \mathbb{A}$, there exists a polynomial of least degree, say, ν with integer coefficients satisfied by α. This is known as the *minimal polynomial* of α. The other roots of the minimal polynomial are called the *conjugates* of α. We shall denote the conjugates of α by $\alpha^{(1)} = \alpha, \alpha^{(2)}, \ldots, \alpha^{(\nu)}$. We denote by $\mathcal{N}(\alpha)$, the *norm* of α which is defined by

$$\mathcal{N}(\alpha) = \alpha^{(1)} \cdots \alpha^{(\nu)}.$$

Observe that for any algebraic number $\alpha \in \mathbb{A}$, there exists a rational integer d such that $d\alpha$ is an algebraic integer. For instance, one may take d to be the leading coefficient of the minimal polynomial satisfied by α over $\mathbb{Q}$. The least such d denoted as $d(\alpha)$ is called the *denominator* of α.

For any polynomial $P(X) \in \mathbb{Q}[X]$, by the *height* of P, denoted as $H(P)$, we mean the maximum of the absolute values of the coefficients of P. By the height of an algebraic number α, denoted as $h(\alpha)$, we mean the height of the minimal polynomial of α. By $\overline{|\alpha|}$, called *house* of α, we mean the maximum of the absolute values of α and its conjugates. By the *size* of α, denoted by $s(\alpha)$, we mean $d(\alpha) + \overline{|\alpha|}$. We will also use the *absolute logarithmic height* of an algebraic number α which we denote by $h°(\alpha)$, and it is defined as

$$h°(\alpha) = \frac{1}{\nu} \log a_0 + \frac{1}{\nu} \sum_{i=1}^{\nu} \log \max(1, |\alpha_i|)$$

where $a_0 > 0$ is the leading coefficient of the minimal polynomial of α.

We know that when α is an algebraic integer, any non-negative integral power of α is a linear combination of $1, \alpha, \ldots, \alpha^{\nu-1}$ with coefficients in $\mathbb{Z}$. The following lemma gives a bound for the coefficients of this linear combination.

Lemma 1.3.1 *Let α be an algebraic integer of degree $\nu \geq 2$ and of height $h(\alpha)$. Let*

$$\alpha^s = \sum_{j=0}^{\nu-1} b_{j,s} \alpha^j, \; s \in \mathbb{N} \cup \{0\}, b_{j,s} \in \mathbb{Z}. \tag{1.3}$$

Then

$$\max_{0 \le j < \nu} |b_{j,s}| \le (2h(\alpha))^s.$$

Proof Note that when $s = 0, 1$, we have, $b_{0,0} = 1, b_{j,0} = 0$ for $j \ge 1$; $b_{0,1} = 0, b_{1,1} = 1, b_{j,1} = 0$ for $j \ge 2$. Hence the lemma is true since $h(\alpha) \ge 1$. We prove the lemma for any s by induction. We assume that the lemma is true for any $t \le s$. Let the minimal polynomial of α be given by

$$X^\nu + A_1 X^{\nu-1} + \cdots + A_\nu.$$

From (1.3), we get

$$\alpha^{s+1} = \alpha \sum_{j=0}^{\nu-1} b_{j,s} \alpha^j = \sum_{j=1}^{\nu} b_{j-1,s} \alpha^j.$$

Also

$$b_{\nu-1,s} \alpha^\nu = -b_{\nu-1,s} \sum_{j=0}^{\nu-1} A_{\nu-j} \alpha^j.$$

Hence

$$\alpha^{s+1} = \sum_{j=0}^{\nu-1} b_{j,s+1} \alpha^j$$

where

$$b_{0,s+1} = -b_{\nu-1,s} A_\nu$$

and

$$b_{j,s+1} = b_{j-1,s} - A_{\nu-j} b_{\nu-1,s} \text{ for } 1 \le j < \nu.$$

Thus

$$\max_{0 \le j < \nu} |b_{j,s+1}| \le 2h(\alpha) \max_{0 \le j < \nu} |b_{j,s}| \text{ for } s \ge 0.$$

By induction hypothesis, applied to the right-hand side of the above inequality, we get

$$\max_{0 \le j < \nu} |b_{j,s+1}| \le (2h(\alpha))^{s+1} \text{ for } s \ge 0.$$

□

The following two lemmas are on size and height of algebraic numbers.

Lemma 1.3.2 *Let α, β be two algebraic number of degree at most ν. Then, we have*

(1) $s(\alpha + \beta) \le s(\alpha) s(\beta)$. Further, if α and β are algebraic integers, then $s(\alpha + \beta) \le s(\alpha) + s(\beta)$;

(2) $s(\alpha\beta) \le s(\alpha) s(\beta)$;

(3) $\dfrac{s(\alpha)^n}{2^{n-1}} \le s(\alpha^n) \le s(\alpha)^n$ *for any natural number n;*

(4) $s(1/\alpha) \le 2s(\alpha)^{2\nu}$.

Proof First note that $d(\alpha + \beta), d(\alpha\beta) \le d(\alpha)d(\beta)$. Also it is clear that $\lceil \alpha + \beta \rceil$, $\lceil \alpha\beta \rceil \le \lceil \alpha \rceil \lceil \beta \rceil$ and hence the assertion of the first part of (1) and (2) follows. If α and β are algebraic integers, then $d(\alpha) = 1 = d(\beta)$. Hence,

$$s(\alpha + \beta) = d(\alpha + \beta) + \lceil \alpha + \beta \rceil \le 1 + \lceil \alpha \rceil + \lceil \beta \rceil \le s(\alpha) + s(\beta).$$

This proves the second part of (1). For any real positive numbers a, b and natural numbers n, by induction on n, we get

$$(a + b)^n \ge a^n + b^n \ge 2\left(\frac{a+b}{2}\right)^n,$$

and hence the assertion (3) follows by taking $a = d(\alpha)$ and $b = \lceil \alpha \rceil$.

Write $\alpha = \alpha'/d(\alpha)$ where α' is an algebraic integer. We have

$$\frac{1}{\alpha} = \frac{d(\alpha)}{\alpha'} = \frac{d(\alpha)\mathcal{N}(\alpha')/\alpha'}{\mathcal{N}(\alpha')}.$$

We see that $\mathcal{N}(\alpha')/\alpha'$ is an algebraic integer and $\mathcal{N}(\alpha')$ is an integer. Thus, we get

$$d(1/\alpha) \le \mathcal{N}(\alpha') = \mathcal{N}(d(\alpha)\alpha) \le d(\alpha)^\nu \lceil \alpha \rceil^\nu \le s(\alpha)^{2\nu}$$

and

$$\lceil 1/\alpha \rceil \le d(\alpha)\lceil \alpha' \rceil^{\nu-1}/\mathcal{N}(\alpha') \le d(\alpha)^\nu \lceil \alpha \rceil^{\nu-1} \le s(\alpha)^{2\nu}.$$

This gives the assertion (4). □

Lemma 1.3.3 *For any non-zero algebraic number α of degree $\le \nu$, we have*

(1) $s(\alpha) \le 2^\nu H(\alpha)^\nu$;

(2) $H(\alpha) \le 2^\nu s(\alpha)^{2\nu}$.

Proof Let α be a non-zero algebraic number of degree $m \le \nu$. Suppose α satisfies $a_0 x^m + a_1 x^{m-1} + \cdots + a_m = 0$ where $\gcd(a_0, \ldots, a_m) = 1$, $a_0 > 0$ and $a_m \ne 0$.

Claim 1 $d(\alpha) \le H(\alpha)$ and $\lceil \alpha \rceil \le 2^{\nu-1} H(\alpha)^\nu$.

First note that since $a_0\alpha$ is an algebraic integer, we get $d(\alpha) \le a_0 \le H(\alpha)$, which proves the first assertion in Claim 1.

Now, we observe the following. If α' is a conjugate of α with $|\alpha'| \le \frac{1}{2}$, then

$$|\alpha'| \ge \frac{1}{2H(\alpha)}. \tag{1.4}$$

For, since α' is a conjugate of α, we get

$$1 \le |a_m| \le |a_0\alpha'^m + \cdots + a_{m-1}\alpha'| \le H(\alpha)|\alpha'| \left(1 + \frac{1}{2} + \frac{1}{2^2} + \cdots\right) = 2H(\alpha)|\alpha'|.$$

Since $\mathcal{N}(\alpha) = \pm a_m/a_0$, we see that $|a_0\mathcal{N}(\alpha)| = |a_m| \le H(\alpha)$ and hence $H(\alpha) \ge |\mathcal{N}(\alpha)|$. Let us suppose that there are r conjugates of α with absolute value $\le 1/2$. Then, we get

$$H(\alpha) \ge \lceil\alpha\rceil \prod_1 |\alpha'| \prod_2 |\alpha''|,$$

where $\prod_1$ runs through all the r conjugates of α whose absolute value $\le 1/2$ and $\prod_2$ runs through the remaining $m - 1 - r$ conjugates of α whose absolute value $> 1/2$. Thus, by (1.4), we get

$$H(\alpha) \ge \lceil\alpha\rceil \prod_1 \frac{1}{2H(\alpha)} \prod_2 \frac{1}{2} = \lceil\alpha\rceil 2^{-(m-1)} H(\alpha)^{-r}.$$

Hence,

$$\lceil\alpha\rceil \le 2^{\nu-1} H(\alpha)^\nu$$

which proves Claim 1.

By Claim 1 and by the definition, we see that

$$s(\alpha) = d(\alpha) + \lceil\alpha\rceil \le H(\alpha) + 2^{\nu-1} H(\alpha)^\nu \le (2H(\alpha))^\nu$$

which proves (1).

In order to prove (2), we first claim the following.

Claim 2 $a_0 \le d^\nu$ where $d = d(\alpha)$.

First note that a_0 divides the $\gcd(a_1 d, \ldots, a_m d^m)$. For, let $1 \le j \le m$ be any integer. Since $d\alpha$ is an algebraic integer,

$$\sum_{1 \le i_1 < \cdots < i_j \le m} (d\alpha_{i_1}) \cdots (d\alpha_{i_j}) = k_j,$$

an integer, where $\alpha_1 = \alpha$ and α_is are other conjugates of α for $i \ge 2$. However

$$k_j = d^j \sum_{1 \le i_1 < \cdots < i_j \le m} \alpha_{i_1} \cdots \alpha_{i_j} = \pm d^j \frac{a_j}{a_0}.$$

Since k_j is an integer, we conclude that a_0 divides $d^j a_j$ for all $j = 1, 2, \ldots, m$ and hence the assertion follows. Since $\gcd(a_1 d, \ldots, a_m d^m)$ divides $d^m \gcd(a_1, \ldots, a_m)$, we get

$$a_0 \text{ divides } d^m \gcd(a_1, \ldots, a_m).$$

Since $\gcd(a_0, a_1, \ldots, a_m) = 1$, we conclude that a_0 divides d^m which in turn proves Claim 2.

Note that $\pm \dfrac{H(\alpha)}{a_0}$ is the j-th elementary symmetric function of α and its conjugates for some $0 \leq j \leq m$. Hence, we get

$$\frac{H(\alpha)}{a_0} \leq \begin{cases} 1; & \text{if } j = 0 \\ 2^\nu \lceil \alpha \rceil^\nu; & \text{if } j > 0. \end{cases}$$

Therefore, by Claim 2, we get

$$H(\alpha) \leq a_0 2^\nu \lceil \alpha \rceil^\nu \leq d^\nu 2^\nu \lceil \alpha \rceil^\nu \leq 2^\nu s(\alpha)^{2\nu},$$

as desired.   □

The above lemma enables us to use size or height as is convenient in any given problem. The next lemma is on polynomials with algebraic coefficients. For any polynomial $P(X) = \alpha_0 + \cdots + \alpha_\nu X^\nu \in \mathbb{A}[X]$, by $\lceil P \rceil$, we mean the maximum of $\lceil \alpha_i \rceil$, $1 \leq i \leq \nu$.

Lemma 1.3.4 *Let* $f_1(X), \ldots, f_t(X) \in \mathbb{A}[X]$. *Then*

$$\left\lceil \prod_{i=1}^{t} f_i \right\rceil \leq \prod_{i=1}^{t} (1 + \deg f_i) \prod_{i=1}^{t} \lceil f_i \rceil.$$

Proof There is no loss of generality in supposing that

$$\deg f_1 \geq \deg f_2 \geq \cdots \geq \deg f_t.$$

The product $f_1 f_2$ is a polynomial, each of whose coefficients is a sum of products of a coefficient of f_1 and of f_2 and the number of such summands is at most $(1 + \deg f_2)$. Hence

$$\lceil f_1 f_2 \rceil \leq (1 + \deg f_2) \lceil f_1 \rceil \lceil f_2 \rceil.$$

Similarly

$$\lceil f_1 f_2 f_3 \rceil \leq (1 + \deg f_3) \lceil f_1 f_2 \rceil \lceil f_3 \rceil \leq (1 + \deg f_3)(1 + \deg f_2) \lceil f_1 \rceil \lceil f_2 \rceil \lceil f_3 \rceil.$$

Now the result follows by induction.   □

The next lemma is on the product of two polynomials in several variables with algebraic coefficients which is a number field generalisation of Lemma 1.2.3. Let $\mathcal{K}$ be an algebraic number field and $\mathcal{O}_{\mathcal{K}}$ be the ring of algebraic integers of $\mathcal{K}$.

Lemma 1.3.5 *Let r and m be positive integers with $1 \leq r < m$. Suppose $F(X_1, \ldots, X_m) \in \mathcal{O}_{\mathcal{K}}[X_1, \ldots, X_m]$; $G(X_1, \ldots, X_r) \in \mathcal{K}[X_1, \ldots, X_r]$; $H(X_{r+1}, \ldots, X_m) \in \mathcal{K}[X_{r+1}, \ldots, X_m]$ such that*

$$F = GH.$$

If γ is any coefficient of F, there is a factorisation $\gamma = \alpha\beta$ in $\mathcal{K}$ such that coefficients of αG and βH are in $\mathcal{O}_{\mathcal{K}}$.

Proof Let the coefficients of G be $\alpha_1, \ldots, \alpha_s$ and those of H be $\beta_1, \ldots, \beta_t$ in some order. Since the variables in G and H are disjoint and the coefficients in F are in $\mathcal{O}_{\mathcal{K}}$, all the products $\alpha_i\beta_1, \ldots, \alpha_i\beta_t$ for $1 \leq i \leq s$ are in $\mathcal{O}_{\mathcal{K}}$ and also $\beta_j\alpha_1, \ldots, \beta_j\alpha_s$ for $1 \leq j \leq t$. But these two sets of numbers are just coefficients of $\alpha_i H$ and $\beta_j G$ which gives the assertion of the lemma. □

Suppose that $\mathcal{K}$ is a number field of degree n. Let $\mathcal{K}$ have r_1 real embeddings and $2r_2$ complex embeddings into $\mathbb{C}$. Thus $n = r_1 + 2r_2$. Let $\sigma_1, \ldots, \sigma_n$ be the n embeddings of $\mathcal{K}$ into $\mathbb{C}$. For any $u \in \mathcal{K}$ we denote $\sigma_i(u)$ by $u^{(i)}$ for $1 \leq i \leq n$ and we suppose that $u^{(i)}$ is real for $1 \leq i \leq r_1$ and $\overline{u^{(r_1+i)}} = u^{(r_1+r_2+i)}$ for $1 \leq i \leq r_2$ are complex. Let $\mathcal{O}_{\mathcal{K}}$ be the ring of integers of $\mathcal{K}$ and let $\mathcal{O}_{\mathcal{K}}^*$ denote the group of units of $\mathcal{O}_{\mathcal{K}}$. Define $\mathcal{N}_{\mathcal{K}}(\alpha)$ and $\overline{|\alpha|}_{\mathcal{K}}$ called the *norm* and *house* of $\alpha \in \mathcal{O}_{\mathcal{K}}$ by

$$\mathcal{N}_{\mathcal{K}/\mathbb{Q}}(\alpha) := \sigma_1(\alpha) \cdots \sigma_n(\alpha), \quad \overline{|\alpha|}_{\mathcal{K}} = \max_{1 \leq i \leq n} |\sigma_i(\alpha)|.$$

It is well known that $\alpha \in \mathcal{O}_{\mathcal{K}}^*$ if and only if $\mathcal{N}_{\mathcal{K}/\mathbb{Q}}(\alpha) = \pm 1$. We state Dirichlet 's unit theorem.

Theorem 1.3.6 *Let $r = r_1 + r_2 - 1$. Define the map*

$$\overline{\log} : \mathcal{O}_{\mathcal{K}}^* \to \mathbb{R}^r : \epsilon \to (\log|\sigma_1(\epsilon)|, \ldots, \log|\sigma_r(\epsilon)|)^T.$$

Then $\overline{\log}$ is a group homomorphism. The kernel of $\overline{\log}$ is the group $U_{\mathcal{K}}$ of roots of unity of $\mathcal{K}$, and this group is finite. The image of $\overline{\log}$ is a lattice in $\mathbb{R}^r$.

Choose units $\eta_1, \ldots, \eta_r$ such that $\overline{\log}(\eta_1), \ldots, \overline{\log}(\eta_r)$ form a basis of the lattice $\overline{\log}(\mathcal{O}_{\mathcal{K}}^*)$. Such $\{\eta_1, \ldots, \eta_r\}$ is called *a system of fundamental units* for $\mathcal{K}$. Then for every $\epsilon \in \mathcal{O}_{\mathcal{K}}^*$, there exist unique integers $b_1, \ldots, b_r$ such that

$$\overline{\log}(\epsilon) = b_1\overline{\log}(\eta_1) + \cdots + b_r\overline{\log}(\eta_r).$$

Hence every $\epsilon \in \mathcal{O}_{\mathcal{K}}^*$ can be expressed uniquely as

$$\epsilon = \zeta\eta_1^{b_1} \cdots \eta_r^{b_r} \text{ with } \zeta \in U_{\mathcal{K}}, b_1, \ldots, b_r \in \mathbb{Z}. \tag{1.5}$$

Further the matrix

$$M := \begin{pmatrix} \log|\sigma_1(\eta_1)| \cdots \log|\sigma_1(\eta_r)| \\ \vdots \\ \log|\sigma_r(\eta_1)| \cdots \log|\sigma_r(\eta_r)| \end{pmatrix}$$

is invertible. We refer to any book on algebraic number theory for these facts. See for instance [3]. With this background we shall derive some consequences in the form of lemmas.

Lemma 1.3.7 *Let $\epsilon \in \mathcal{O}_\mathcal{K}^*$ be written as in (1.5). Then there exists an effectively computable number $c_{1.1} = c_{1.1}(\mathcal{K}, \eta_1, \ldots, \eta_r) > 0$ such that*

$$\max\left(|b_1|, \ldots, |b_r|\right) \leq c_{1.1} \log\left(\overline{\lceil\epsilon\rceil}_\mathcal{K}\right).$$

Proof Note that

$$\overline{\log}(\epsilon) = M\mathbf{b}$$

where $\mathbf{b} = (b_1, \ldots, b_r)^T$. Hence $\mathbf{b} = M^{-1}\overline{\log}(\epsilon)$. Thus if $M^{-1} = (a_{ij})$, we obtain

$$b_i = \sum_{j=1}^{r} a_{ij} \log|\sigma_j(\epsilon)| \text{ for } 1 \leq i \leq r.$$

We have $\overline{\lceil\epsilon\rceil} \geq 1$ and since $\mathcal{N}_{\mathcal{K}/\mathbb{Q}}(\epsilon) = \pm 1$, we get $|\sigma_j(\epsilon)| \geq (\overline{\lceil\epsilon\rceil})^{1-n}$ for $1 \leq j \leq n$. Hence

$$|\log|\sigma_j(\epsilon)|| \leq n\log(\overline{\lceil\epsilon\rceil}) \text{ for } 1 \leq j \leq n.$$

Thus

$$\max_{1 \leq i \leq r}|b_i| \leq \left(\max_{1 \leq i \leq r}\sum_{j=1}^{r}|a_{ij}|\right) n\log(\overline{\lceil\epsilon\rceil}) = c_{1.1}\log(\overline{\lceil\epsilon\rceil}).$$

$\square$

In the next lemma we shall show that for every $\alpha \in \mathcal{O}_\mathcal{K}\backslash\{0\}$ there exists an $\epsilon \in \mathcal{O}_\mathcal{K}^*$ such that $\epsilon\alpha$ and all its conjugates have about the same absolute value. This enables us to see that $\overline{\lceil\epsilon\alpha\rceil}_\mathcal{K}$ is about the n-th root of $|\mathcal{N}_{\mathcal{K}/\mathbb{Q}}(\alpha)|$.

Lemma 1.3.8 *Let $\alpha \in \mathcal{O}_\mathcal{K}\backslash\{0\}$. Then there exist an effectively computable number $c_{1.2} = c_{1.2}(\mathcal{K}, \eta_1, \ldots, \eta_r) > 0$ and an $\epsilon \in \mathcal{O}_\mathcal{K}^*$ such that*

$$c_{1.2}^{-1}|\mathcal{N}_{\mathcal{K}/\mathbb{Q}}(\alpha)|^{1/n} \leq \overline{\lceil\epsilon\alpha\rceil}_\mathcal{K} \leq c_{1.2}|\mathcal{N}_{\mathcal{K}/\mathbb{Q}}(\alpha)|^{1/n}.$$

Proof We use the following fact (See for instance, [2]). Let $\mathcal{L}$ be a lattice in $\mathbb{R}^r$. Then for every $\mathbf{X} = (x_1, \ldots, x_r) \in \mathbb{R}^r$ there is a point $\mathbf{u} \in \mathcal{L}$ such that $||\mathbf{u} - \mathbf{X}||_2 \leq c_{1.3}(\mathcal{L})$, where $||X||_2 = \sqrt{x_1^2 + \cdots + x_r^2}$. Now we take $\mathcal{L} = \overline{\log}(\mathcal{O}_\mathcal{K}^*)$. Then for every $\mathbf{X} \in \mathbb{R}^r$, there exists an $\epsilon \in \mathcal{O}_\mathcal{K}^*$ with

$$||\mathbf{X} - \overline{\log}(\epsilon)||_2 \leq c_{1.3}(\mathcal{K}, \eta_1, \ldots, \eta_r) = c_{1.3}.$$

This implies that

$$||x_i - \log|\sigma_i(\epsilon)|| \le c_{1.3} \text{ for } 1 \le i \le r.$$

Now we take $\mathbf{X}$ with $x_i = -\log|\sigma_i(\alpha)| + \frac{1}{n}\log|\mathcal{N}_{\mathcal{K}/\mathbb{Q}}(\alpha)|$, $1 \le i \le r$. Then there exists $\epsilon \in \mathcal{O}_{\mathcal{K}}^*$ such that

$$\left|\log|\sigma_i(\epsilon)| + \log|\sigma_i(\alpha)| - \frac{1}{n}\log\left|\mathcal{N}_{\mathcal{K}/\mathbb{Q}}(\alpha)\right|\right| \le c_{1.3} \text{ for } 1 \le i \le r$$

or

$$\left|\log|\sigma_i(\epsilon\alpha)| - \frac{1}{n}\log\left|\mathcal{N}_{\mathcal{K}/\mathbb{Q}}(\alpha)\right|\right| \le c_{1.3} \text{ for } 1 \le i \le r.$$

Now let

$$\xi_i = \log|\sigma_i(\epsilon\alpha)| - \frac{1}{n}\log\left|\mathcal{N}_{\mathcal{K}/\mathbb{Q}}(\alpha)\right| \text{ for } 1 \le i \le n,$$

and

$$e_i = \begin{cases} 1 \text{ for } 1 \le i \le r_1 \\ 2 \text{ for } r_1 + 1 \le i \le r_1 + r_2 = r + 1. \end{cases}$$

Since $\log|\sigma_{r_1+r_2+i}(\epsilon\alpha)| = \log|\sigma_{r_1+i}(\epsilon\alpha)|$ for $1 \le i \le r_2$ and

$$\sum_{i=1}^{n}\log|\sigma_i(\epsilon\alpha)| = \log|\mathcal{N}_{\mathcal{K}/\mathbb{Q}}(\epsilon\alpha)| = \log|\mathcal{N}_{\mathcal{K}/\mathbb{Q}}(\alpha)|$$

we have

$$\xi_{r_1+r_2+i} = \xi_{r_1+i} \text{ for } 1 \le i \le r_2 \text{ and } \sum_{i=1}^{r_1+r_2} e_i\xi_i = 0.$$

Hence

$$|\xi_{r_1+r_2}| \le e_{r_1+r_2}^{-1}\sum_{i=1}^{r} e_i|\xi_i| \le c_{1.3}e_{r_1+r_2}^{-1}\sum_{i=1}^{r} e_i = c_{1.4}, \text{ say.}$$

Thus

$$\left|\log|\sigma_i(\epsilon\alpha)| - \frac{1}{n}\log\left|\mathcal{N}_{\mathcal{K}/\mathbb{Q}}(\alpha)\right|\right| = |\xi_i| \le \max(c_{1.3}, c_{1.4}) \text{ for } 1 \le i \le n.$$

Taking i with $\overline{|\epsilon\alpha|}_{\mathcal{K}} = |\sigma_i(\epsilon\alpha)|$ we get the lemma with $c_{1.2} = e^{\max(c_{1.3}, c_{1.4})}$. $\qquad\square$

As a consequence of the above lemma, we derive the following corollary.

Corollary 1.3.9 *Let $\alpha \in \mathcal{O}_{\mathcal{K}}\backslash\{0\}$ be given. Then there exists a finite set of divisors $\{\gamma_1, \ldots, \gamma_m\}$ of α in $\mathcal{O}_{\mathcal{K}}$ such that every divisor β of α in $\mathcal{O}_{\mathcal{K}}$ can be written as $\beta = \epsilon^{(i)}\gamma_i$ for some i with $1 \le i \le m$ and $\epsilon^{(i)} \in \mathcal{O}_{\mathcal{K}}^*$.*

Proof Let β be a divisor of α. That is, there exists $\gamma \in \mathcal{O}_\mathcal{K}$ such that $\alpha = \gamma\beta$. Then $\mathcal{N}_{\mathcal{K}/\mathbb{Q}}(\beta)$ divides $\mathcal{N}_{\mathcal{K}/\mathbb{Q}}(\alpha)$. By Lemma 1.3.8, there exists an $\epsilon \in \mathcal{O}_\mathcal{K}^*$ such that

$$\overline{|\epsilon\beta|}_\mathcal{K} \le c_{1.2}|\mathcal{N}_{\mathcal{K}/\mathbb{Q}}(\beta)|^{1/n} \le c_{1.2}|\mathcal{N}_{\mathcal{K}/\mathbb{Q}}(\alpha)|^{1/n}.$$

But there are only finitely many $\gamma \in \mathcal{O}_\mathcal{K}$ with degree at most n and

$$\overline{|\gamma|}_\mathcal{K} \le c_{1.2}|\mathcal{N}_{\mathcal{K}/\mathbb{Q}}(\alpha)|^{1/n}.$$

These can be effectively determined and each one of them is checked if it divides α in $\mathcal{O}_\mathcal{K}$. Thus we get a finite set $\{\gamma_1, \ldots, \gamma_m\}$ of divisors of α and any divisor β of α is of the form $\epsilon\gamma_i$ for some $1 \le i \le m$. $\square$

1.4 Linear Independence of Functions

Let $f_1(X), \ldots, f_m(X)$ be m polynomials in $\mathbb{Z}[X]$. Let D be the determinant given by

$$D = D(f_1, \ldots, f_m) := ||f_\lambda^{(k)}(X)||, 1 \le \lambda \le m; 0 \le k < m,$$

where $f^{(k)}(X)$ denotes the k-th derivative of $f(X)$. This is called the *Wronskian* of $f_1, \ldots, f_m$. The following lemma gives a criterion for linear dependence of $f_1, \ldots, f_m$ in terms of the Wronskian .

Lemma 1.4.1 *The polynomials* $f_1(X), \ldots, f_m(X)$ *are linearly dependent over* $\mathbb{Q}$ *if and only if* $D \equiv 0$.

Proof Suppose $f_1, \ldots, f_m$ are linearly dependent over $\mathbb{Q}$, i.e. there exists a relation

$$\sum_{\lambda=1}^{m} c_\lambda f_\lambda(X) = 0 \text{ with } c_\lambda \in \mathbb{Q} \text{ and not all zero.}$$

Differentiating the above relation $m - 1$ times, we get

$$\sum_{\lambda=1}^{m} c_\lambda f_\lambda^{(k)}(X) = 0, \text{ for all } 0 \le k < m.$$

Since $(c_\lambda)_\lambda$ is a non-zero solution to the system, we conclude that the determinant D must be 0.

Suppose now $D \equiv 0$. We prove the lemma by induction on m. That is, we assume that if the Wronskian of any $m - 1$ functions is identically zero, then the $m - 1$ functions are linearly dependent. Then we show that it is true for any m functions.

We may assume that $f_1(X) \not\equiv 0$ since, otherwise, the lemma is true. Let us consider an interval I in which $f_1(X) \neq 0$. Note that if $g(X)$ is a function differentiable

$m - 1$ times, then

$$D(g f_1, \ldots, g f_m) = g^m D(f_1, \ldots, f_m).$$

Take $g(X) = 1/f_1(X)$ for $X \in I$. Then

$$D\left(1, \frac{f_2(X)}{f_1(X)}, \ldots, \frac{f_m(X)}{f_1(X)}\right) = 0$$

which implies

$$D\left(\frac{d}{dX}(f_2/f_1), \ldots, \frac{d}{dX}(f_m/f_1)\right) = 0.$$

Hence, by the induction hypothesis, the functions $\frac{d}{dX}(f_2/f_1), \ldots, \frac{d}{dX}(f_m/f_1)$ are linearly dependent over $\mathbb{Q}$. So there exist constants $c_2, \ldots, c_m \in \mathbb{Q}$, not all zero, such that

$$c_2 \frac{d}{dX}(f_2/f_1) + \cdots + c_m \frac{d}{dX}(f_m/f_1) = 0.$$

Hence there exists a constant $c_1 \in \mathbb{Q}$ such that

$$c_2 f_2(X) + \cdots + c_m f_m(X) = -c_1 f_1(X).$$

Since this dependence relation holds for all $X \in I$ and f_is are polynomials, this holds for all values of X. $\qquad\qquad\Box$

We saw in Lemma 1.4.1 that linear independence of polynomials of single variable is valid if and only if their Wronskian does not vanish. For polynomials of several variables the situation is not simple since there are several partial derivatives to consider. We proceed as follows. Let $f_0, \ldots, f_{\ell-1}$ be functions of $X_1, \ldots, X_m$. Let $\Delta_0, \ldots, \Delta_\mu, \ldots, \Delta_{\ell-1}$ be differential operators of the form

$$\frac{1}{j_1! \cdots j_m!} \left(\frac{\partial}{\partial X_1}\right)^{j_1} \cdots \left(\frac{\partial}{\partial X_m}\right)^{j_m},$$

such that $j_1 + \cdots + j_m$ of Δ_μ does not exceed μ for $0 \le \mu \le \ell - 1$. Then the function

$$G(z_1, \ldots, z_m) = \begin{vmatrix} \Delta_0 f_0 & \Delta_0 f_1 & \cdots & \Delta_0 f_{\ell-1} \\ \Delta_1 f_0 & \Delta_1 f_1 & \cdots & \Delta_1 f_{\ell-1} \\ \cdots & \cdots & \cdots & \\ \Delta_{\ell-1} f_0 & \Delta_{\ell-1} f_1 & \cdots & \Delta_{\ell-1} f_{\ell-1} \end{vmatrix}$$

is called a *generalised Wronskian* of $f_0, \ldots, f_{\ell-1}$. When $m = \ell = 1$, we get $\mu = 0$ and $\Delta_0 = f_0$. In all other cases, there are several Δ_μs for each μ, and hence more than one generalised Wronskian. When $m = 1$, i.e. in the case of functions of one variable, the ordinary Wronskian is that generalised Wronskian for which the order of Δ_μ is exactly μ for $0 \le \mu \le \ell - 1$. Roth [4] used the notion of generalised Wronskian in an ingenious way to construct a polynomial $P(X_1, \ldots, X_m)$ in m variables, m large, so that $P(p_1/q_1, \ldots, p_m/q_m) \ne 0$ where p_i/q_i are rationals very close to a given algebraic number α. In the next two lemmas, such a construction is shown. These will be used in Chap. 7.

Lemma 1.4.2 *Let $f_0, \ldots, f_{\ell-1}$ be ℓ polynomials in $\mathbb{Q}[X_1, \ldots, X_m]$ for which every generalised Wronskian $G(X_1, \ldots, X_m)$ vanishes identically. Then $f_0, \ldots, f_{\ell-1}$ are linearly dependent over $\mathbb{Q}$.*

Proof Proof is by contradiction. Suppose the polynomials $f_0, \ldots, f_{\ell-1}$ are linearly independent. Assume that each f_ν is of degree $\le k$ for X_i, $1 \le i \le m$. Thus we can write

$$f_\nu(X_1, \ldots, X_m) = \sum_{k_1=0}^{k-1} \cdots \sum_{k_m=0}^{k-1} b_\nu(k_1, \ldots, k_m) X_1^{k_1} \ldots X_m^{k_m}, 0 \le \nu \le \ell - 1.$$

We claim that $f_\nu(t, t^k, t^{k^2}, \ldots, t^{k^{m-1}})$ are linearly independent as polynomials in t. Suppose not. Then there exist c_ν, $0 \le \nu \le \ell - 1$, not all zero, such that

$$\sum_{\nu=0}^{\ell-1} c_\nu \sum_{k_1=0}^{k-1} \cdots \sum_{k_m=0}^{k-1} b_\nu(k_1, \ldots, k_m) t^{k_1+k_2 k+\cdots+k_m k^{m-1}} = 0$$

or

$$\sum_{k_1=0}^{k-1} \cdots \sum_{k_m=0}^{k-1} \left(\sum_{\nu=0}^{\ell-1} c_\nu b_\nu(k_1, \ldots, k_m) \right) t^{k_1+k_2 k+\cdots+k_m k^{m-1}} = 0.$$

From the uniqueness of representation of an integer to the base k for each set of exponents $(k_1, \ldots, k_m)$, it follows that

$$\sum_{\nu=0}^{\ell-1} c_\nu b_\nu(k_1, \ldots, k_m) = 0,$$

for all tuples $(k_1, \ldots, k_m)$ which in turn gives

$$\sum_{\nu=0}^{\ell-1} c_\nu f_\nu(X_1, \ldots, X_m) = 0$$

which contradicts our assumption that f_is are linearly independent. This proves the claim.

By Lemma 1.4.1, we know then that the Wronskian

$$W(t) = \det\left(\frac{1}{\mu!}\left(\frac{d}{dt}\right)^{\mu} f_{\nu}\left(t, t^k, \ldots, t^{k^{m-1}}\right)\right), \mu, \nu = 0, \ldots, \ell - 1$$

does not vanish identically. By a differentiation formula,

$$\frac{d}{dt} f_{\nu}(t, \ldots, t^{k^{m-1}}) = \sum_{j=1}^{m} \frac{\partial}{\partial X_j} f_{\nu}(X_1, \ldots, X_m)|_{(t, \ldots, t^{k^{m-1}})} \frac{d}{dt}(t^{k^{j-1}})$$

and by induction on μ we get an operator identity

$$\left(\frac{d}{dt}\right)^{\mu} = \phi_1(t)\Delta^{(1)} + \cdots + \phi_r(t)\Delta^{(r)}$$

where $\Delta^{(1)}, \ldots, \Delta^{(r)}$ are differential operators of orders not exceeding μ and r depends only on μ and m. Further $\phi_1, \ldots, \phi_r$ are polynomials with rational coefficients. We use this in the expression for $W(t)$ and writing the resulting determinant as a sum of other determinants, we get an expression for $W(t)$ as

$$W(t) = \psi_1(t)G_1(t, \ldots, t^{k^{m-1}}) + \cdots + \psi_s(t)G_s(t, \ldots, t^{k^{m-1}})$$

in which $\psi_1, \ldots, \psi_s$ are polynomials and $G_1, \ldots, G_s$ are generalised Wronskians of $f_0, \ldots, f_{\ell-1}$. Since $W(t)$ does not vanish identically, there is an i for which $G_i(t, \ldots, t^{k^{m-1}})$ is not identically zero and a fortiori $G_i(X_1, \ldots, X_m)$ is not identically zero which contradicts the assumption. □

In the next lemma we show that given any polynomial in m variables $X_1, \ldots, X_m$ with rational coefficients, we can construct a polynomial with integral coefficients which can be decomposed as a product of a polynomial with integral coefficients in $m - 1$ variables $X_1, \ldots, X_{m-1}$ and another polynomial with integral coefficients in the variable X_m.

Lemma 1.4.3 *Let $R(X_1, \ldots, X_m) \in \mathbb{Q}[X_1, \ldots, X_m]$ in $m \geq 2$ variables with $\deg_{X_j} R \leq r_j, 1 \leq j \leq m$ and $H(R) \leq B$. Then there are integers β, ℓ with $1 \leq \ell \leq r_m + 1$, and differential operators $\Delta_0, \ldots, \Delta_{\ell-1}$ on the variables $X_1, \ldots, X_{m-1}$ of orders at most $0, \ldots, \ell - 1$, respectively, such that if*

$$F(X_1, \ldots, X_m) = \beta \det\left(\Delta_{\mu}\frac{1}{\nu!}\left(\frac{\partial}{\partial X_m}\right)^{\nu} R\right), 0 \leq \mu, \nu \leq \ell - 1,$$

then

(a) $F \in \mathbb{Z}[X_1, \ldots, X_m]$ and $F \not\equiv 0$
(b) F can be written as

$$F(X_1, \ldots, X_m) = U(X_1, \ldots, X_{m-1})V(X_m)$$

where $U \in \mathbb{Z}[X_1, \ldots, X_{m-1}]$ with $\deg_{X_j} U \leq \ell r_j$, $1 \leq j \leq m-1$ and $V(X_m) \in \mathbb{Z}[X_m]$ with $\deg_{X_m} V \leq \ell r_m$.

(c) $H(F) \leq ((r_1 + 1) \cdots (r_m + 1))^{2\ell} 2^{2\ell(r_1 + \cdots + r_m)} (\ell!)^2 B^{2\ell}$.

Proof Write R as a polynomial in X_m as

$$R(X_1, \ldots, X_m) = \sum_{i=0}^{r_m} S_i(X_1, \ldots, X_{m-1}) X_m^i.$$

The polynomials S_i need not be linearly independent. Therefore there exists an integer ℓ such that a set of polynomials $\xi_\nu(X_1, \ldots, X_{m-1})$, $0 \leq \nu < \ell$ is a maximal set of linearly independent polynomials among the S_i. Thus $1 \leq \ell \leq r_m + 1$. Then there are constants $\beta_{\nu i} \in \mathbb{Q}$ such that for $0 \leq i \leq r_m$ such that

$$S_i(X_1, \ldots, X_{m-1}) = \sum_{\nu=0}^{\ell-1} \beta_{\nu i} \xi_\nu(X_1, \ldots, X_{m-1}). \tag{1.6}$$

Put

$$\phi_\nu(X_m) = \sum_{i=0}^{r_m} \beta_{\nu i} X_m^i, \, 0 \leq \nu < \ell.$$

Then

$$R(X_1, \ldots, X_m) = \sum_{\nu=0}^{\ell-1} \xi_\nu(X_1, \ldots, X_{m-1}) \phi_\nu(X_m).$$

We claim that the polynomials ϕ_ν, $0 \leq \nu < \ell$ are linearly independent. Suppose not. Then there are constants $\delta_0, \ldots, \delta_{\ell-1}$ such that

$$\delta_0 \phi_0(X_m) + \cdots + \delta_{\ell-1} \phi_{\ell-1}(X_m) = 0.$$

The coefficient of each power of X_m must be zero. This gives

$$\delta_0 \beta_{0i} + \cdots + \delta_{\ell-1} \beta_{(\ell-1)i} = 0, \, 0 \leq i \leq r_m. \tag{1.7}$$

We know $\{\xi_0, \ldots, \xi_{\ell-1}\} \subseteq \{S_0, \ldots, S_{r_m}\}$. For a fixed ν_0, $0 \leq \nu_0 < \ell$, choose h_0 so that

$$S_{h_0}(X_1, \ldots, X_{m-1}) = \xi_{\nu_0}(X_1, \ldots, X_{m-1}).$$

Then by (1.6)

$$\beta_{\nu h_0} = \begin{cases} 1 & \text{if } \nu = \nu_0 \\ 0 & \text{if } \nu \neq \nu_0. \end{cases}$$

Hence by (1.7), with $i = h_0$, we obtain $\delta_{\nu_0} = 0$. Since ν_0 is arbitrary, every $\delta_i = 0$ which proves the claim.

Let $W(X_m)$ be the Wronskian of $\phi_0, \ldots, \phi_{\ell-1}$. Thus $W(X_m) \in \mathbb{Q}[X_m]$ and $\not\equiv 0$, by Lemma 1.4.1. Since $\xi_0, \ldots, \xi_{\ell-1}$ are linearly independent, by Lemma 1.4.2, there exists $G(X_0, \ldots, X_{m-1})$, some generalised Wronskian of $\xi_0, \ldots, \xi_{\ell-1}$ which is not identically zero. Then

$$W(X_m) = \det\left(\frac{1}{\mu!}\left(\frac{d}{dX_m}\right)^\mu \phi_\nu(X_m)\right)$$

and

$$G(X_1, \ldots, X_{m-1}) = \det(\Delta_\mu \xi_\nu(X_1, \ldots, X_{m-1}))$$

for $0 \le \mu, \nu < \ell$. Here $\Delta_0, \ldots, \Delta_{\ell-1}$ are differential operators on $X_1, \ldots, X_{m-1}$ of orders at most $0, \ldots, \ell - 1$, respectively. Thus

$$GW = \det\left(\sum_{\rho=0}^{\ell-1} \Delta_\rho \frac{1}{\nu!}\left(\frac{\partial}{\partial X_m}\right)^\nu \phi_\rho(X_m)\xi_\mu(X_1, \ldots, X_{m-1})\right),$$

or

$$GW = \det\left(\Delta_\mu \frac{1}{\nu!}\left(\frac{\partial}{\partial X_m}\right)^\nu R\right).$$

Since W is a determinant of order ℓ whose elements are polynomials in X_m of degrees at most r_m, we get $\deg W \le \ell r_m$. Similarly G is of degree at most ℓr_j in X_j for $1 \le j < m$. In the expression for GW, we can write R as the sum of $(r_1 + 1)\cdots(r_m + 1)$ terms of the form

$$\alpha_{s_1\ldots s_m} X_1^{s_1} \cdots X_m^{s_m}$$

with $|\alpha_{s_1\ldots s_m}| \le B$. The determinant can then be written as a sum of

$$((r_1 + 1)\cdots(r_m + 1))^\ell$$

new determinants, each having entries of the form

$$\alpha_{s_1\ldots s_m} \Delta_\mu \frac{1}{\nu!}\left(\frac{\partial}{\partial X_m}\right)^\nu X_1^{s_1} \cdots X_m^{s_m} = a\alpha_{s_1\ldots s_m} X_1^{t_1} \cdots X_m^{t_m}$$

in which $t_j \le s_j$ for $1 \le j \le m$. Here

$$a = \binom{s_1}{t_1} \cdots \binom{s_m}{t_m} \le 2^{s_1+\cdots+s_m} \le 2^{r_1+\cdots+r_m}.$$

Thus the absolute value of the coefficient of $X_1^{t_1} \cdots X_m^{t_m}$ does not exceed

$$2^{r_1 + \cdots + r_m} B.$$

There are $\ell!$ terms in the expansion of each new determinant, and the absolute value of the coefficient of a typical term in this expansion does not exceed

$$2^{\ell(r_1 + \cdots + r_m)} B^\ell.$$

Hence

$$H(GW) \le ((r_1 + 1) \cdots (r_m + 1))^\ell \ell! 2^{\ell(r_1 + \cdots + r_m)} B^\ell.$$

Coefficients of GW are in $\mathbb{Q}$. Hence by Lemma 1.3.5 there is a $\beta \in \mathbb{Q}$ such that $\beta = \beta_1 \beta_2$ in $\mathbb{Q}$ with $\beta_1 G = U$ and $\beta_2 W = V$ have coefficients in $\mathbb{Z}$ and $\beta GW = F = UV$. Further

$$0 < H(F) \le (H(GW))^2 \le ((r_1 + 1) \cdots (r_m + 1))^{2\ell} (\ell!)^2 2^{2\ell(r_1 + \cdots + r_m)} B^{2\ell}. \qquad \square$$

Remark

Suppose $R(X_1, \ldots, X_m) \in \mathbb{Z}[X_1, \ldots, X_m]$, then in the above lemma, $\beta = 1$ and $F = GW$ with $G[X_1, \ldots, X_{m-1}] \in \mathbb{Z}[X_1, \ldots, X_{m-1}]$ and $W(X_m) \in \mathbb{Z}[X_m]$. Hence

$$H(F) = H(GW) \le ((r_1 + 1) \cdots (r_m + 1))^\ell \ell! 2^{r_1 + \cdots + r_m} B^\ell. \qquad (1.8)$$

Exercise

1. Let $f(z)$ be a complex valued function not identically zero. Suppose $Z(f)$, the set of zeros of $f(z)$ is infinite. Then show that z and $f(z)$ are algebraically independent over $\mathbb{Q}$. (Hint: Suppose z and $f(z)$ satisfy $a_0(z) f(z)^\nu + \cdots + a_\nu(z) = 0$. For $z \notin Z(f)$, divide by $f(z)^\nu$ and let $z \to z_0$ where $z_0 \in Z(f)$ to conclude $a_0(z) \equiv 0$. Apply induction to conclude the proof.)
2. For any algebraic number α, show that $|\alpha| \le h(\alpha) + 1$.
3. In Lemma 1.3.8, compute $c_{1.3}(\mathcal{L})$ when $\mathcal{L}$ is the lattice generated by $(1, 0)$ and $(0, 1)$ in $\mathbb{R}^2$.

Notes

As noted earlier, it is a well-known fact from complex analysis that an entire function of order $\le \rho$ has $O(R^\rho)$ number of zeros in a disc of radius R for large R. This has been generalised to exponential polynomials by Tijdeman [5] in 1973, and his result was very precise. This became a powerful tool in transcendence theory. In 2014, Senthil Kumar et al. [6], gave an equivalent criterion for a generalisation to Lemma 1.1.2.

References

1. L.V. Ahlfors, *Complex Analysis*, 3rd edn. (McCraw-Hill, New York, 1979)
2. J.W.S. Cassels, *An Introduction to Diophantine Approximation*, Cambridge Tracts (1957), 166 pp.
3. S. Lang, *Algebraic Number Theory*, vol. 110, 2nd edn., Graduate Texts in Mathematics (Springer, New York, 1994)
4. K.F. Roth, Rational approximations to algebraic numbers. Mathematika **2**, 1–20 (1955); Corrigendum **2**, 168 (1955)
5. R. Tijdeman, An auxiliary result in the theory of transcendental numbers. J. Number Theory **5**, 80–94 (1973)
6. K. Senthil Kumar, R. Thangadurai, M. Waldschmidt, Liouville numbers and Schanuel's conjecture. Arch. Math. (Basel) **102**, 59–70 (2014)

Chapter 2
Early Transcendence Results from Nineteenth Century

A beam in darkness, let it grow

—Tennyson

The transcendental number theory was originated by Liouville in 1844 when he showed a method to construct transcendental numbers. For instance,

$$\sum_{n=0}^{\infty} \frac{1}{2^{n!}}$$

is transcendental. On the other hand, the transcendence of classical constants like e and π remained elusive although the irrationality of these numbers was known long back. In 1737, Euler found the continued fraction expansion of e as

$$e = [2, 1, 2, 1, 1, 4, 1, 1, 6, 1, 1, \ldots].$$

To prove the irrationality, it is enough to use the fact that this continued fraction is infinite. This was noticed by Lambert in 1766. He used continued fractions to prove the irrationality of π as well. We refer to [1] for these historical facts.

The transcendence of e was first proved by Hermite in 1873 by using very different ideas and applying the approximation of analytic functions by rational functions. See Sects. 2.1 and 2.2. By modifying Hermite's method, Lindemann proved the transcendence of π in 1882. This settles an ancient problem of the impossibility of squaring a circle. See Sect. 2.3. Further he extended the ideas of Hermite to prove more general results on the linear independence of the values of the exponential function. His proof was later simplified by Weierstrass in 1885. See Sects. 2.4 and 2.5. The theorem of Hermite–Lindemann–Weierstrass leads to transcendental results

© Springer Nature Singapore Pte Ltd. 2020
S. Natarajan and R. Thangadurai, *Pillars of Transcendental Number Theory*,
https://doi.org/10.1007/978-981-15-4155-1_2

on exponential function and trigonometric functions at algebraic values and also on a very first algebraic independence result on a set of numbers. See Sect. 2.6.

2.1 Functional Identity of Hermite

Hermite used the following identity to prove the transcendence of e.

Lemma 2.1.1 *Let $\phi(X) \in \mathbb{Q}[X]$ be a polynomial of height $H(\phi)$ and degree d. Then*

$$I(\phi, x) := e^x \int_0^x \phi(t)e^{-t}dt = F(0)e^x - F(x) \tag{2.1}$$

where $F(x) = \sum_{k \geq 0} \phi^{(k)}(x)$ in which $\phi^{(k)}(x)$ denotes the k-th derivative of $\phi(x)$. Further,

$$|I(\phi, x)| \leq dH(\phi)e^x x^{d+1}. \tag{2.2}$$

Proof The identity (2.1) follows from integration by parts as

$$I(\phi, x) = -e^x \int_0^x \phi(t)d(e^{-t}) = -\phi(x) + e^x\phi(0) - e^x \int_0^x \phi^{(1)}(t)e^{-t}dt.$$

Consider

$$|I(\phi, x)| = \left| \int_0^x \phi(t)e^{x-t}dt \right| \leq xe^x \max_{t \in (0,x)} |\phi(t)|.$$

The upper bound follows since

$$\max_{t \in (0,x)} |\phi(t)| \leq \begin{cases} dH(\phi) \text{ if } x \leq 1 \\ dH(\phi)x^d \text{ if } x > 1. \end{cases}$$

$\square$

Remark 1 We specialise $\phi(X)$ as $f(X) = X^{N-1}((X-1)\cdots(X-n))^N$. Then since f has zero of order $N-1$ at $X = 0$ and of order N at $X = k$ for any integer $1 \leq k \leq n$, we observe that

$$f^{(j)}(0) \equiv 0 \pmod{N!}$$

for $j \neq N - 1$ and

$$f^{(N-1)}(0) = (N-1)!(-1)^{nN}(n!)^N.$$

Further,

$$f^{(j)}(k) \equiv 0 \pmod{N!}$$

for $1 \le k \le n$ and $j \ge 0$. Thus we get

$$
\begin{cases}
F(0) \equiv (N-1)!(-1)^{nN}(n!)^N \pmod{N!} \\
F(k) \equiv 0 \pmod{N!} \text{ for } 1 \le k \le n.
\end{cases}
\tag{2.3}
$$

Thus for any integer k, we have

$$
|I(f, k)| \ge (N-1)!
$$

provided $N > n$ and N is a prime, as $(N-1)!(-1)^{nN}(n!)^N \not\equiv 0 \pmod{N}$ and $\equiv 0 \pmod{(N-1)!}$.

We shall use Hermite's identity in the complex domain as below.

Lemma 2.1.2 *Let $a_0, \ldots, a_m; \alpha_0, \ldots, \alpha_m \in \mathbb{C}$ be such that*

$$
a_0 e^{\alpha_0} + \cdots + a_m e^{\alpha_m} = 0.
\tag{2.4}
$$

Let

$$
\phi(z) = (z - \alpha_0)^n (z - \alpha_1)^{n+1} \cdots (z - \alpha_m)^{n+1}
$$

and

$$
g(z) = \frac{1}{n!} \sum_{\ell \ge n} \phi^{(\ell)}(z).
$$

Then

$$
|a_0 g(\alpha_0) + \cdots + a_m g(\alpha_m)| \le \frac{c_{2.1}^{n+1}}{n!}
$$

where $c_{2.1}$ depends only on a_i's, α_i's and is independent of n.

Proof As before, let

$$
F(z) = \sum_{\ell \ge 0} \phi^{(\ell)}(z).
$$

Then (2.1) holds with the path of integration being the line joining 0 and z. Thus

$$
F(\alpha_k) = \sum_{\ell \ge n} \phi^{(\ell)}(\alpha_k) = n! g(\alpha_k), 0 \le k \le m.
$$

Also by (2.1) and (2.4), we have

$$\sum_{k=0}^{m} a_k g(\alpha_k) = \frac{1}{n!} \sum_{k=0}^{m} a_k F(\alpha_k) - \frac{1}{n!} \sum_{k=0}^{m} a_k F(0)e^{\alpha_k} + \frac{1}{n!} \sum_{k=0}^{m} a_k F(0)e^{\alpha_k}$$

$$= -\frac{1}{n!} \sum_{k=0}^{m} a_k \int_{0}^{\alpha_k} e^{\alpha_k - z} \phi(z) dz.$$

Let $r = \max_{0 \le k \le m} (|\alpha_k|)$. Then

$$\left| \sum_{k=0}^{m} a_k \int_{0}^{\alpha_k} e^{\alpha_k - z} \phi(z) dz \right| \le \sum_{k=0}^{m} |a_k| e^{|\alpha_k|} (2r)^{(m+1)(n+1)} \le c_{2.1}^{n+1}$$

where one may take $c_{2.1} = e^r (2r)^{(m+1)} \left(1 + \sum_{k=0}^{m} |a_k| \right)$. $\square$

2.2 e Is Transcendental

Proof Suppose e satisfies

$$P(e) = 0$$

for some $P(X) = a_n X^n + a_{n-1} X^{n-1} + \cdots + a_0 \in \mathbb{Z}[X]$ with $a_n \ne 0$. We use Lemma 2.1.1 and Remark 1 that follows. $\square$

Choice of $\phi(X)$

For any integer $N > 0$, let

$$\phi(X) = X^{N-1}((X-1) \cdots (X-n))^N \text{ and } F(z) = \sum_{\ell \ge 0} \phi^{(\ell)}(z).$$

Consider

$$S = \sum_{k=0}^{n} a_k F(k).$$

Lower Bound for $|S|$

By (2.3),

$$S \equiv a_0 (N-1)! (-1)^{nN} (n!)^N \pmod{N!}.$$

Taking N to be a prime exceeding n and $|a_0|$ we see that

$$S \ne 0 \text{ and } |S| \ge (N-1)! \tag{2.5}$$

Upper Bound for $|S|$

Note that

$$S = \sum_{k=0}^{n} a_k I(\phi, k) - F(0) \sum_{k=0}^{n} a_k e^k = \sum_{k=0}^{n} a_k I(\phi, k)$$

since $P(e) = 0$. Also $d = \deg(\phi) \le (n+1)N$. Observe that

$$|\phi(X)| \le (|X| + n)^{(n+1)N}.$$

Hence

$$H = H(\phi) = \max_{0 \le k \le (n+1)N} \binom{(n+1)N}{k} n^k \le (2n)^{(n+1)N}.$$

We use the estimate from (2.2) to get

$$|S| \le \sum_{k=0}^{n} |a_k|(n+1)N(2n)^{(n+1)N} e^k k^{(n+1)N+1}$$

$$\le (n+1)N(2n)^{(n+1)N} e^n n^{(n+1)N+1} \left(\sum_{k=0}^{n} |a_k| \right)$$

$$\le c_{2.2}^N$$

where $c_{2.2}$ depends on n and a_k's and is independent of N. For instance, one may take $c_{2.2}$ as $e(2n)^{2(n+2)}(\sum |a_k|)^{1/n}$.

Final Contradiction

The upper bound above for $|S|$ contradicts the lower bound in (2.5) for N sufficiently large.   □

2.3 π Is Transcendental

We begin with the well-known theorem on symmetric functions.

Lemma 2.3.1 *Let $P(t_1, \ldots, t_n) \in \mathbb{Z}[t_1, \ldots, t_n]$ be a polynomial and of degree r in each of t_i. Suppose P is symmetric in $t_1, \ldots, t_n$. Then P is equal to a polynomial with coefficients in $\mathbb{Z}$, of total degree r in the elementary symmetric functions*

$$\sum_{1 \le i \le n} t_i, \quad \sum_{1 \le i < j \le n} t_i t_j, \ldots, \prod_{i=1}^{n} t_i.$$

As a consequence, we obtain the following lemma.

Lemma 2.3.2 *Suppose $\phi(X) \in \mathbb{Z}[X]$ is a polynomial of degree d with leading coefficient $a > 0$. Further let $\alpha_1, \ldots, \alpha_d$ be the roots of $\phi(X) = 0$. Then any symmetric rational integral polynomial in $a\alpha_1, \ldots, a\alpha_d$ is a rational integer.*

Proof Let

$$\phi(X) = aX^d + a_1 X^{d-1} + \cdots + a_d.$$

Then any symmetric rational integral polynomial in $a\alpha_1, \ldots, a\alpha_d$ is a rational integral polynomial in

$$\sum_{1 \le i \le d} a\alpha_i, \quad \sum_{1 \le i < j \le d} a\alpha_i a\alpha_j, \ldots, \prod_{i=1}^{d} a\alpha_i$$

by Lemma 2.3.1 and hence a rational integral polynomial in $a_1, \ldots, a_d$ showing that it is a rational integer.

Now we prove the transcendence of π. Suppose π is algebraic. Then πi is also algebraic. Let $\alpha_1 = \pi i$ be of degree n satisfying the polynomial

$$aX^n + a_1 X^{n-1} + \cdots + a_n = 0.$$

Since $1 + e^{\alpha_1} = 0$, we have

$$(1 + e^{\alpha_1})(1 + e^{\alpha_2}) \cdots (1 + e^{\alpha_n}) = 0$$

where $\alpha_2, \ldots, \alpha_n$ are conjugates of α_1. The product on the left-hand side can be expanded as

$$\sum \exp\left(\sum_{j=1}^{n} \epsilon_j \alpha_j\right)$$

where the first sum is over all tuples $(\epsilon_1, \ldots, \epsilon_n)$ with $\epsilon_j \in \{0, 1\}$ for $1 \le j \le n$. Thus each summand is of the form e^θ, and there are 2^n such summands. Assume that exactly d of the θs is non-zero. Then the above sum becomes

$$2^n - d + e^{\theta_1} + \cdots + e^{\theta_d} = 0.$$

Hence with $I(\phi, x)$ as defined in Lemma 2.1.1, we get

$$S := (2^n - d)I(\phi, 0) + I(\phi, \theta_1) + \cdots + I(\phi, \theta_d)$$

$$= -(2^n - d) \sum_{j \ge 0} \phi^{(j)}(0) - \sum_{j \ge 0} \sum_{k=1}^{d} \phi^{(j)}(\theta_k),$$

$\qquad\qquad\qquad\qquad\qquad\qquad\qquad\qquad\qquad\qquad\qquad\qquad\qquad\qquad\qquad\square$

Choice of $\phi(X)$

Now we specialise $\phi(X)$ as

$$\phi(X) = a^{dp} X^{p-1}((X - \theta_1) \cdots (X - \theta_d))^p$$

where p is a prime with

$$p > \max(a, 2^n - d, a^d \theta_1 \ldots \theta_d).$$

Note that

$$a^d X^{2^n - d}(X - \theta_1) \cdots (X - \theta_d) = \prod_{j=1}^{n} \prod_{\epsilon_j=0}^{1} \left(aX - \sum_{i=1}^{n} \epsilon_i a\alpha_i\right).$$

The polynomial on the right-hand side of the above equality is a polynomial with coefficients which are symmetric polynomials in $a\alpha_1, \ldots, a\alpha_n$ with rational integer coefficients and hence are rational integers by Lemma 2.3.2. So by the above identity and the definition of $\phi(X)$, it follows that

$$\phi(X) \in \mathbb{Z}[X].$$

Lower Bound for $|S|$

As seen earlier, we have

$$\phi^{(p-1)}(0) = (p - 1)!(-a)^{dp}(\theta_1 \cdots \theta_d)^p \in \mathbb{Z}$$

and the integer on the left-hand side is divisible by $(p - 1)!$ but $p!$ does not divide it by the choice of p. Further

$$\phi^{(j)}(0) \in \mathbb{Z} \text{ and } \equiv 0 \pmod{p!} \text{ for } j \neq p - 1$$

and

$$\phi^{(j)}(\theta_k) \in \mathbb{Z} \text{ and } \equiv 0 \pmod{p!} \text{ for } j \geq 0.$$

Thus S is an integer divisible by $(p - 1)!$ but $p!$ does not divide S. Hence S is non-zero and

$$|S| \geq (p - 1)!$$

Upper Bound for $|S|$

Note that

$$|S| \leq |I(\phi, \theta_1)| + \cdots + |I(\phi, \theta_d)|$$

$$= \sum_{k=1}^{d} \left| \int_0^{\theta_k} e^{\theta_k - t} \phi(t) dt \right|$$

$$\leq \sum_{k=1}^{d} e^{|\theta_k|} \int_0^{\theta_k} |\phi(t)| dt$$

$$\leq \sum_{k=1}^{d} e^{|\theta_k|} |\theta_k| \hat{\phi}(|\theta_k|)$$

$$\leq c_{2.3}^p.$$

where $\hat{\phi}(X) = a^{dp} |X|^{p-1} (|X| + |\theta_1|)^p \cdots (|X| + |\theta_d|)^p$ and $c_{2.3}$ is a positive number depending on a, d and θs and independent of p. For instance, $c_{2.3}$ can be taken as $d(2a)^d |\theta_0|^{d+2} e^{|\theta_0|}$ where $|\theta_0| = \max(|\theta_1|, \ldots, |\theta_d|)$.

Final Contradiction

Comparing the upper and lower bounds for $|S|$ we get

$$(p-1)! \leq |S| \leq c_{2.3}^p$$

which is not possible if p is sufficiently large. For instance, p can be taken $> c_{2.3}$.

$\square$

2.4 A Lemma from Galois Theory

Lemma 2.4.1 *Let $\mathcal{K}$ be a normal extension of $\mathbb{Q}$ with $[\mathcal{K} : \mathbb{Q}] = \nu$ and $\sigma_1, \ldots, \sigma_\nu$ be the automorphisms of $\mathcal{K}$. Let $A(z) \in \mathcal{K}[[z]]$ be a power series. Then*

$$B(z) = \prod_{i=1}^{\nu} \sigma_i(A(z)) \in \mathbb{Q}[[z]].$$

Proof Observe that for any $A(z) = \sum_{k \geq 0} \gamma_k z^k \in \mathcal{K}[[z]]$, and any automorphism σ_i of $\mathcal{K}$, we have

$$\sigma_i(A(z)) = \sum_{k \geq 0} \sigma_i(\gamma_k) z^k.$$

Hence for any automorphism σ, we have

$$\sigma(B(z)) = \sigma \prod_{i=1}^{\nu} (\sigma_i(A(z))) = \prod_{i=1}^{\nu} (\sigma \sigma_i(A(z)) = \prod_{j=1}^{\nu} \sigma_j(A(z)) = B(z).$$

Hence $B(z) \in \mathbb{Q}[[z]]$.

$\square$

2.5 Theorem of Hermite–Lindemann–Weierstrass

Theorem 2.5.1 *Let* $\alpha_0, \ldots, \alpha_m$ *be distinct algebraic numbers and* $a_0, \ldots, a_m$ *be non-zero algebraic numbers. Then*

$$\sum_{i=0}^{m} a_i e^{\alpha_i} \neq 0.$$

In other words, if $\alpha_0, \ldots, \alpha_m$ are distinct algebraic numbers, then $e^{\alpha_0}, \ldots, e^{\alpha_m}$ are linearly independent over $\mathbb{A}$. The proof is similar to the proof of the transcendence of e, but more sophisticated as we will now deal with a normal extension of $\mathbb{Q}$.

Proof We assume that

$$\sum_{i=0}^{m} a_i e^{\alpha_i} = 0. \tag{2.6}$$

$\square$

Application of Lemma 2.4.1

Let us consider the function

$$A(z) = \sum_{i=0}^{m} a_i e^{\alpha_i z}, z \in \mathbb{C}.$$

Since α_is are distinct, $A(z) \not\equiv 0$. For, suppose $\sum_{i=0}^{m} a_i e^{\alpha_i z} = 0$ for all z. In particular,

we get $\sum_{i=0}^{m} a_i e^{\alpha_i j} = 0$ for all $j = 1, 2, \ldots, m+1$. Since the matrix $(e^{j \alpha_i})$ has non-zero determinant as α_is are distinct, we get $a_i = 0$ for all i, which is a contradiction.

Let $\mathcal{K}$ be the normal extension of $\mathbb{Q}$ containing a_is and α_is and let

$$[\mathcal{K} : \mathbb{Q}] = \nu.$$

Since $A(1) = 0$ by (2.6), we see that $\sigma(A(1)) = 0$ for all automorphisms of $\mathcal{K}$ and hence we may replace $A(z)$ by $B(z)$ in Lemma 2.4.1 to assume that $A(z) \in \mathbb{Q}[[z]]$.

Application of Lemma 2.1.2

Since $A(1) = 0$, we may multiply $A(1)$ by a common denominator of $a_0, \ldots, a_m$ to assume that $a_i \in \mathcal{O}_{\mathcal{K}}$, for $0 \leq i \leq m$. Let $d \in \mathbb{Z}$ be the common denominator of $\alpha_i, 0 \leq i \leq m$. Further let n be a large integer to be chosen later and let

$$f(z) = (z - \alpha_0)^n (z - \alpha_1)^{n+1} \cdots (z - \alpha_m)^{n+1};$$

$$h(z) = (z - d\alpha_0)^n (z - d\alpha_1)^{n+1} \cdots (z - d\alpha_m)^{n+1}$$

and

$$g(z) = \frac{1}{n!} \sum_{\ell \geq n} f^{(\ell)}(z).$$

Then

$$f(z) \in \mathcal{K}[z]; \, h(z) \in \mathcal{O}_{\mathcal{K}}[z]$$

and since

$$d^{m(n+1)} f(z) = d^{-n} h(dz)$$

we get

$$d^{m(n+1)} \frac{f^{(\ell)}(\alpha_j)}{\ell!} = d^{\ell-n} \frac{h^{(\ell)}(d\alpha_j)}{\ell!} \in \mathcal{O}_{\mathcal{K}}, \text{ for } \ell \geq n.$$

Let

$$I = d^{m(n+1)} \sum_{j=0}^{m} a_j g(\alpha_j).$$

An Upper Bound for $|I|$

We have

$$I = \sum_{j=0}^{m} a_j \sum_{\ell \geq n} d^{m(n+1)} \frac{1}{n!} f^{(\ell)}(\alpha_j)$$

$$= \frac{d^{m(n+1)} a_0 f^{(n)}(\alpha_0)}{n!} + (n+1) \sum_{j=0}^{m} \sum_{\ell \geq n+1} a_j \frac{\ell!}{(n+1)!} d^{m(n+1)} \frac{f^{(\ell)}(\alpha_j)}{\ell!}.$$

Note that

$$\frac{f^{(n)}(\alpha_0)}{n!} = \prod_{j=1}^{m} (\alpha_0 - \alpha_j)^{n+1}.$$

Thus we get from the above expression for I that

$$I = a_0 \prod_{j=1}^{m} (d\alpha_0 - d\alpha_j)^{n+1} + (n+1)J \tag{2.7}$$

where $J \in \mathcal{O}_{\mathcal{K}}$. Suppose $I = 0$. Then from (2.7), we see that

$$(n+1) \text{ divides } \mathcal{N}(a_0) \prod_{j=1}^{m} \mathcal{N}(d\alpha_0 - d\alpha_j)^{n+1}.$$

By taking

$$n = k \left| \mathcal{N}(d\alpha_0 - d\alpha_j) \right|, \tag{2.8}$$

for some $1 \le j \le m$ with $k \ge |\mathcal{N}(a_0)|$ we see that the above property does not hold. Hence $I \ne 0$. By Lemma 2.1.2, with $\phi = f$ we conclude that

$$0 < |I| < \frac{c_{2.4}^{n+1}}{n!} \tag{2.9}$$

where $c_{2.4} = d^m c_{2.1}$ with $c_{2.1}$ as given in Lemma 2.1.2. Hence $c_{2.4}$ depends only on a_is and α_is.

Equation (2.9) is Valid for $|\sigma(I)|$

Let $\sigma_1, \ldots, \sigma_\nu$ be the automorphisms of $\mathcal{K}$. Since $A(z) \in \mathbb{Q}[[z]]$, we have $\sigma_i(A(z)) = A(z)$ for $1 \le i \le \nu$. Hence $A(1) = 0$ implies

$$\sigma_i(A(1)) = \sum_{\ell=0}^{m} \sigma_i(a_\ell) e^{\sigma_i(\alpha_\ell)} = 0$$

for any $1 \le i \le \nu$. Thus replacing f, g, I by $\sigma_i f, \sigma_i g, \sigma_i I$, we get as above that

$$|\sigma_i(I)| \le \frac{c_{2.5}^{n+1}}{n!}, \quad 1 \le i \le \nu$$

where $c_{2.5}$ depends only on a_is and α_is.

Final Step

We therefore have

$$|\mathcal{N}(I)| = \left| \prod_{i=1}^{\nu} \sigma_i(I) \right| \le \frac{c_{2.5}^{\nu(n+1)}}{(n!)^\nu}.$$

Now $\mathcal{N}(I)$ is a non-zero integer since $I \in \mathcal{O}_\mathcal{K}$ and $I \ne 0$. Hence the above inequality is impossible since the right-hand side goes to 0 as $n \to \infty$ which is valid by taking k in (2.8) sufficiently large. $\qquad \square$

2.6 Applications of Theorem 2.5.1

It is immediate that e is transcendental by taking $\alpha_i = i$ for $0 \le i \le m$. Also e^α is transcendental for any non-zero algebraic α. As a consequence, we get $\log \beta$ is transcendental for any algebraic β different from 0 and 1. Further π is transcendental since $e^{2\pi i} = 1$. The trigonometric functions $\sin \alpha, \cos \alpha, \tan \alpha$ are transcendental for any non-zero algebraic α. The proof of this fact is easy and left to the reader. An important consequence is the following corollary on algebraic independence of some numbers. This was the first result on the algebraic independence of a set of numbers.

Corollary 2.6.1 *Suppose* $\alpha_1, \ldots, \alpha_r$ *are algebraic numbers which are linearly independent over* $\mathbb{Q}$. *Then* $e^{\alpha_1}, \ldots, e^{\alpha_r}$ *are algebraically independent over* $\mathbb{Q}$.

Proof Suppose $e^{\alpha_1}, \ldots, e^{\alpha_r}$ are algebraically dependent over $\mathbb{Q}$. Then there exists a polynomial $P(x_1, \ldots, x_r) \in \mathbb{Q}[x_1, \ldots, x_r]$ such that $P(e^{\alpha_1}, \ldots, e^{\alpha_r}) = 0$. This can be rewritten as

$$\sum_{(k_1, \ldots, k_r)} a_{k_1, \ldots, k_r} e^{k_1 \alpha_1 + \cdots + k_r \alpha_r} = 0 \qquad (2.10)$$

with $a_{k_1, \ldots, k_r} \in \mathbb{Q}$ and not all zero. Note that $k_1 \alpha_1 + \cdots + k_r \alpha_r$ are all distinct since $\alpha_1, \ldots, \alpha_r$ are linearly independent over $\mathbb{Q}$. Thus (2.10) contradicts Hermite–Lindemann–Weierstrass Theorem. $\qquad\qquad\square$

Exercise

1. Using the series expansion of e, show that e is irrational. Will the same argument work for e^2?
2. Show that $1, e, e^{-1}$ are linearly independent over $\mathbb{Q}$. (Hint: Use the series expansion of e^{-1}.)
3. Show that at least one of $e + \pi$ and $e\pi$ is transcendental.
4. Prove that $e^{\sqrt{2}}$ and $e^{\sqrt{3}}$ are algebraically independent over $\mathbb{A}$.
5. Let $F(X_1, \ldots, X_n) \in \mathbb{A}[X_1, \ldots, X_n]$ be a non-zero polynomial. Suppose $\theta_1, \ldots, \theta_n$ are algebraically independent numbers. Show that $F(\theta_1, \ldots, \theta_n)$ is transcendental.
6. Let $c_0, c_1, \ldots,$ be a periodic sequence of algebraic numbers. Show that the values of the series

$$f(z) = \sum c_n \frac{z^n}{n!}$$

at any non-zero algebraic number z is transcendental.

Notes

Here is an example of a more general series than that of e. Let $\{b_n\}_{n \geq 1}$ be a sequence of integers with $b_n \geq 2$ for all integers $n \geq 1$. Let $\{c_n\}$ be a sequence of integers with $c_n \in \{1, 2, \ldots, b-1\}$. Suppose $c_n/b_n \to 0$ as $n \to \infty$. Then the following series (known as *Cantor series*)

$$\sum_{n=1}^{\infty} \frac{c_n}{b_1 \cdots b_n}$$

is irrational. In particular, taking $c_n = 1, b_n = n$ we get the irrationality of e. More information on e and π can be found in [2, 3].

In 1929, Siegel introduced E-functions which are generalisations of exponential function. Results on algebraic independence of values of E-functions at algebraic points were proved by Shidlovsky in 1955. See [2] and [4]. A very powerful conjecture which covers many of the algebraic independence results involving the values of the exponential function is as follows:

Schanuel's conjecture: *Suppose* $\alpha_1, \ldots, \alpha_n$ *are complex numbers which are linearly independent over* $\mathbb{Q}$. *Then at least n numbers from the set* $\{\alpha_1, \ldots, \alpha_n, e^{\alpha_1}, \ldots, e^{\alpha_n}\}$ *are algebraically independent.*

For several consequences of this conjecture we refer to [5].

References

1. N.I. Fel'dman, Y.V. Nesterenko, *Number Theory IV: Transcendental Numbers*, vol. 44, Encyclopaedia of Mathematical Sciences (Springer, Berlin, 1991)
2. Y.V. Nesterenko, *Algebraic Independence*, vol. 14 (Tata Institute of Fundamental Research Publications, 2008), 157 pp
3. I. Niven, *Irrational Numbers* (Cambridge University, Cambridge), 164 pp
4. C.L. Siegel, *Transcendental Numbers* (Princeton University Press, Princeton, 1949)
5. M.R. Murty, P. Rath, *Transcendental Numbers* (Springer, Berlin, 2014), 217 pp

Chapter 3
Theorem of Gelfond and Schneider

The powers of the mind are like the rays of the sun when they are
concentrated they illumine
—Vivekananda

The twentieth century turned out to be a golden era for transcendental number theory. Several important results were proved during this period. In August 1900, at the second International Congress of Mathematicians (ICM) held in Paris, Hilbert proposed a list of 23 problems which, according to him, will have profound value for the progress of mathematical sciences. The seventh problem in the list was titled, "irrationality and transcendence of certain numbers". In this, he raised the question whether an irrational logarithm of an algebraic number to an algebraic base is transcendental. This can be asked differently as whether an irrational quotient of natural logarithms of algebraic numbers is transcendental or whether α^β is transcendental for any algebraic number $\neq 0, 1$ and any algebraic irrational β. The question about properties of $\log_\alpha \beta$ with α, β rational was stated by Euler. He supposed that they are transcendental with exceptions like $\log_2 4$ but no proofs were given. Although this problem has similarity with the Hermite–Lindemann–Weierstrass theorem, Hilbert felt that it is extraordinarily difficult. He was of the view that proof of the irrationality of $2^{\sqrt{2}}$ belongs to the more distant future than the proof of Riemann Hypothesis or Fermat's Last Theorem. He was mistaken.

The first progress towards Hilbert's seventh problem was made by Gelfond in 1929. He showed that α^β is transcendental for any algebraic number $\alpha \neq 0, 1$ and any *imaginary quadratic irrational β*. In particular this proves the transcendence of $e^\pi = (-1)^{-i}$. In 1930, Kuzmin extended to *real quadratic irrational β*. Four years later, in 1934, Hilbert's seventh problem was completely solved by Gelfond and Schneider, independently by different methods although there were many common features. The main difference lies in the construction of auxiliary function which vanishes at certain points. While Gelfond constructed an auxiliary function that has

© Springer Nature Singapore Pte Ltd. 2020
S. Natarajan and R. Thangadurai, *Pillars of Transcendental Number Theory*,
https://doi.org/10.1007/978-981-15-4155-1_3

zeros with high multiplicity, Schneider's auxiliary function has simple zeros but they are two dimensional, i.e. they depend on two integer parameters.

Here we give the proof of Gelfond–Schneider theorem as presented by Siegel.

Theorem 3.0.1 *Let α, β be algebraic numbers, $\alpha \neq 0, 1$ and β irrational. Then α^β is transcendental.*

3.1 Lemmas on Linear Equations

The following two lemmas deal with solutions of a system of linear equations. It is well known from linear algebra that a given system of w number of homogeneous linear equations in v variables with integer coefficients has infinitely many integer solutions, if $v > w$. The following lemma gives a bound for a non-trivial solution.

Lemma 3.1.1 *Let*

$$y_j := a_{j1}x_1 + \cdots + a_{jv}x_v = 0 \; for \; 1 \leq j \leq w$$

be w linear equations in v variables $x_1, \ldots, x_v$ with $a_{ji} \in \mathbb{Z}$, $1 \leq j \leq w$; $1 \leq i \leq v$ and let $v > w$. Assume that $|a_{ji}| \leq A$. Then there exist $x_1, \ldots, x_v \in \mathbb{Z}$, not all zero, satisfying the equations such that

$$|x_i| \leq (vA)^{\frac{w}{v-w}}.$$

Proof Let $s_j = \sum_{a_{ji} \geq 0} a_{ji}$ and $t_j = -\sum_{a_{ji} < 0} a_{ji}$ for any j with $1 \leq j \leq w$. Then $s_j + t_j \leq vA$. For all $1 \leq i \leq v$, let $0 \leq x_i \leq X$ where $X \geq 1$ will be chosen later. Then

$$-Xt_j \leq y_j \leq Xs_j, 1 \leq j \leq w. \tag{3.1}$$

Let $\psi : \mathbb{R}^v \to \mathbb{R}^w$ be given by

$$\psi(x_1, \ldots, x_v) = (y_1, \ldots, y_w).$$

The number of integral tuples $(x_1, \ldots, x_v)$ is $(X + 1)^v$ while the number of integral tuples $(y_1, \ldots, y_w)$ satisfying (3.1) is at most $(1 + X(s_j + t_j))^w \leq (1 + XvA)^w$. Thus if

$$(X + 1)^v > (1 + XvA)^w, \tag{3.2}$$

there exist two distinct tuples $(x_1', \ldots, x_v')$ and $(x_1'', \ldots, x_v'')$ such that

$$\psi(x_1', \ldots, x_v') = \psi(x_1'', \ldots, x_v'') = (y_1^{(0)}, \ldots, y_w^{(0)}).$$

Hence $(x_1, \ldots, x_v) = ((x_1' - x_1''), \ldots, (x_v' - x_v''))$ satisfies the linear system with $x_i \in \mathbb{Z}$, not all zero and

$$|x_i| \leq X.$$

We choose

$$X = [(vA)^{\frac{w}{v-w}}]$$

where $[x]$ denotes the integral part of x. Then (3.2) is satisfied and the proof is complete.

Lemma 3.1.2 *Suppose* $\{\gamma_1, \ldots, \gamma_v\}$ *is a* $\mathbb{Z}$*-basis for* $\mathcal{O}_\mathcal{K}$. *Let* $\alpha \in \mathcal{O}_\mathcal{K}$ *be written as*

$$\alpha = a_1\gamma_1 + \cdots + a_v\gamma_v, a_i \in \mathbb{Z}. \tag{3.3}$$

Then

$$|a_i| \leq c_{3.1}\overline{|\alpha|}$$

where $c_{3.1} > 0$ *depends only on* $\mathcal{K}$.

Proof Taking conjugate on both sides of (3.3), we have

$$\alpha^{(i)} = a_1\gamma_1^{(i)} + \cdots + a_v\gamma_v^{(i)}, 1 \leq i \leq v.$$

Then by Cramer's rule,

$$a_k = \frac{\det A_k}{\det(\gamma_j^{(i)})} \tag{3.4}$$

where A_k is the matrix $(\gamma_j^{(i)})$ with its kth column replaced by $(\alpha^{(1)}, \ldots, \alpha^{(v)})^T$. It is well known that $\det(\gamma_j^{(i)}) \neq 0$. Further $\det A_k$ can be written as a linear expression in $\alpha^{(1)}, \ldots, \alpha^{(v)}$ with coefficients defined in terms of $\gamma_j^{(i)}$. Taking absolute values on both sides of (3.4) we get the result.

We now give an analogue of Lemma 3.1.1 when the coefficients of the linear equations are in $\mathcal{O}_\mathcal{K}$.

Lemma 3.1.3 *Let* $\mathcal{K}$ *be a number field of degree* $v > 1$.

$$y_j := a_{j1}x_1 + \cdots + a_{jv}x_v = 0 \text{ for } 1 \leq j \leq w \tag{3.5}$$

be w *linear equations in* v *variables* $x_1, \ldots, x_v$ *with* $a_{ji} \in \mathcal{O}_\mathcal{K}, 1 \leq i \leq v; 1 \leq j \leq w$ *and* $v > w$. *Assume that* $\overline{|a_{ji}|} \leq A$. *Then the following assertions hold.*

(i) *There exist* $x_1, \ldots, x_v \in \mathcal{O}_\mathcal{K}$, *not all zero, satisfying the equations and positive numbers* $c_{3.2}$ *and* $c_{3.3}$ *depending only on* $\mathcal{K}$ *such that*

$$\overline{|x_i|} \leq c_{3.2}(c_{3.3}vA)^{\frac{w}{v-w}}.$$

(ii) There exist $x'_1, \ldots, x'_v \in \mathbb{Z}$, not all zero, satisfying the equations such that

$$|x'_i| \leq 1 + (2vA)^{\frac{wv(v+1)}{2v - wv(v+1)}}, \ 1 \leq i \leq v$$

provided $2v > wv(v+1)$ and $A \geq 1$.

Proof of (i) We denote by $c_{3.4}, c_{3.5}$ positive numbers depending only on $\mathcal{K}$.

Let $\gamma_1, \ldots, \gamma_v$ be a basis for $\mathcal{O}_\mathcal{K}$. If $x_1, \ldots, x_v \in \mathcal{O}_\mathcal{K}$ are solutions of (3.5), then write

$$x_s = \xi_{s1}\gamma_1 + \cdots + \xi_{sv}\gamma_v, \ 1 \leq s \leq v \tag{3.6}$$

with all $\xi_{si} \in \mathbb{Z}$. Further write

$$a_{js}\gamma_r = b_{jsr1}\gamma_1 + \cdots + b_{jsrv}\gamma_v,$$

for $1 \leq j \leq w, 1 \leq s \leq v, 1 \leq r \leq v$ with all $b_{jsri} \in \mathbb{Z}$. Then for $1 \leq j \leq w$,

$$
\begin{aligned}
0 &= \sum_{s=1}^{v} a_{js} x_s = \sum_{s=1}^{v} a_{js} \sum_{r=1}^{v} \xi_{sr}\gamma_r \\
&= \sum_{r=1}^{v} \sum_{s=1}^{v} \xi_{sr} \sum_{u=1}^{v} b_{jsru}\gamma_u \\
&= \sum_{u=1}^{v} \left(\sum_{r=1}^{v} \sum_{s=1}^{v} b_{jsru}\xi_{sr} \right) \gamma_u.
\end{aligned}
$$

Since $\gamma_1, \ldots, \gamma_v$ is a basis for $\mathcal{K}$,

$$\sum_{r=1}^{v} \sum_{s=1}^{v} b_{jsru}\xi_{sr} = 0 \tag{3.7}$$

for $1 \leq j \leq w$ and $1 \leq u \leq v$. This gives wv equations in vv variables ξ_{sr}. Further by Lemma 3.1.2 and $\lceil ab \rceil \leq \lceil a \rceil \lceil b \rceil$, we get,

$$|b_{jsru}| \leq c_{3.1}\lceil a_{js}\gamma_r \rceil \leq c_{3.1} A \max_{1 \leq r \leq v} \lceil \gamma_r \rceil \leq c_{3.4} A.$$

Hence by Lemma 3.1.1, the system of Eq. (3.7) has a non-trivial solution in $\mathbb{Z}$ satisfying

$$|\xi_{sr}| < (c_{3.4}Avv)^{w/(v-w)}, \ 1 \leq s \leq v, 1 \leq r \leq v.$$

Hence from (3.6), we get

$$\overline{|x_s|} \le |\xi_{s1}| \overline{|\gamma_1|} + \cdots + |\xi_{sv}| \overline{|\gamma_v|}$$
$$\le v(c_{3.4} A v v)^{w/(v-w)} \max_{1 \le r \le v} \overline{|\gamma_r|}$$
$$\le c_{3.5}(c_{3.4} A v v)^{w/(v-w)}.$$

This completes the proof of (i) of the lemma.

Proof of (ii) Let $X \ge 2$ and Y be natural numbers. Let

$$I_X = \{(x_1, \ldots, x_v) : |x_i| \le X, 1 \le i \le v\}$$

and

$$J_Y = \{(y_1, \ldots, y_w) : \overline{|y_j|} \le Y, 1 \le j \le w\}.$$

Note that $\overline{|y_j|} \le v A X$ for any $(x_1, \ldots, x_v) \in I_X$. Hence there is a mapping from I_X to J_{vAX}. We have

$$|I_X| \le (2X + 1)^v.$$

Now $y_j \in \mathcal{O}_K$ and hence satisfies an equation of the form

$$y_j^v + b_1 y_j^{v-1} + \cdots + b_v = 0, \ b_i \in \mathbb{Z}, 1 \le i \le v.$$

In fact, for $1 \le i \le v$, b_i is the ith symmetric function in y_j and its conjugates. So

$$|b_i| \le \binom{v}{i}(vAX)^i.$$

Hence the number of possible equations satisfied by y_j is at most

$$\prod_{i=1}^{v}\left(2\binom{v}{i}(vAX)^i + 1\right),$$

and each such equation has at most v possible values for y_j. Thus

$$|J_{vAX}| \le \left\{v \prod_{i=1}^{v}\left(2\binom{v}{i}(vAX)^i + 1\right)\right\}^w = v^w 2^{wv}(vAX)^{w\frac{v(v+1)}{2}}\left(\prod_{i=1}^{v}\binom{v}{i} + \frac{1}{2(vAX)^i}\right)^w.$$

Since $vAX \ge 2$, using the arithmetic–geometric means, the right-hand side of the above inequality can be estimated as

$$v^w 2^{wv}(vAX)^{wv(v+1)/2}\prod_{i=1}^{v}\left(\binom{v}{i} + \frac{1}{2^{i+1}}\right)^w \le v^w 2^{wv}\left(\frac{2^v - 1/2}{v}\right)^{wv}(vAX)^{wv(v+1)/2}.$$

Thus we get

$$|J_{vAX}| < 2^{wv(v+1)}(vAX)^{wv(v+1)/2}.$$

Let

$$(2X+1)^v \geq 2^{wv(v+1)}(vAX)^{wv(v+1)/2} = (4vAX)^{wv(v+1)/2}. \qquad (3.8)$$

Arguing as in Lemma 3.1.1 there exist $x'_1, \ldots, x'_v \in \mathbb{Z}$, not all zero, satisfying the given system of linear equations with $|x'_i| \leq 2X$. Define λ as

$$\lambda^{v-wv(v+1)/2} = (2vA)^{wv(v+1)/2}$$

and take X satisfying $\lambda - 1 \leq 2X < \lambda + 1$. Then

$$\begin{aligned}(4vAX)^{wv(v+1)/2} &= (2vA)^{wv(v+1)/2}(2X)^{wv(v+1)/2} \\ &< (2X+1)^{v-wv(v+1)/2}(2X)^{wv(v+1)/2} \\ &< (2X+1)^v.\end{aligned}$$

Hence (3.8) is satisfied. Also since $2X \leq \lambda + 1$, we get the assertion of the lemma.

$\square$

3.2 Proof of Gelfond–Schneider Theorem 3.0.1

Assume that $\alpha, \beta, \omega = \alpha^\beta$ are all algebraic. Let $\mathcal{K}$ be the number field containing these numbers with $[\mathcal{K} : \mathbb{Q}] = \nu$.

Choice of Parameters
Let

$$m = 2\nu + 2; \ t = u^2; \ 2m|t; \ n = t/(2m).$$

Thus

$$u = c_{3.6}\sqrt{n}$$

where $c_{3.6} = 2\sqrt{\nu + 1}$. Here and henceforth, we denote by $c_{3.6}, c_{3.7}, \ldots$ positive numbers depending only on α, β and independent of n. Let

$$\theta_{au+b} = (a + b\beta) \log \alpha, a, b \in \mathbb{Z} \text{ and } 0 \leq a \leq u - 1, 1 \leq b \leq u.$$

There are $t = u^2$ such θs and let them be labelled as $\theta_1, \ldots, \theta_t$. Note that θs are all distinct since β is irrational.

Auxiliary Polynomial
Put

$$R(z) = \eta_1 e^{\theta_1 z} + \cdots + \eta_t e^{\theta_t z} \qquad (3.9)$$

where $\eta_1, \ldots, \eta_t$ are variables to be determined such that

$$(\log \alpha)^{-k} R^{(k)}(s) = 0 \text{ for } 0 \leq k \leq n-1 \text{ and } 1 \leq s \leq m. \tag{3.10}$$

This is a set of mn equations in $t(= 2mn)$ variables $\eta_1, \ldots, \eta_t$.

Coefficients of $\eta_1, \ldots, \eta_k$ in (3.9) are in $\mathcal{K}$
A typical coefficient is of the form

$$(\log \alpha)^{-k} \theta_i^k e^{\theta_i s} = (a + b\beta)^k e^{s(a+b\beta)\log \alpha} = (a + b\beta)^k \alpha^{sa} \omega^{sb}$$

and hence in $\mathcal{K}$ by our assumption on ω. Let $c_{3.7}$ be a common denominator of α, β and ω. Then

$$c_{3.7}\alpha, c_{3.7}\beta, c_{3.7}\omega \in \mathcal{O}_{\mathcal{K}}.$$

Hence

$$c_{3.7}^{n-1+2mu}(\log \alpha)^{-k} R^{(k)}(s) \tag{3.11}$$

have coefficients in $\mathcal{O}_{\mathcal{K}}$. Also *house* of the coefficients is bounded by

$$c_{3.7}^{n-1+2mu}\left(u + u\lceil\beta\rceil\right)^{n-1} \lceil\alpha\rceil^{mu} \lceil\omega\rceil^{mu} \leq c_{3.8}^n u^{n-1} \leq c_{3.9}^n t^{(n-1)/2} \leq c_{3.10}^n n^{(n-1)/2}$$

since $u = c_{3.6}\sqrt{n}$.

Application of Lemma 3.1.3
By Lemma 3.1.3 (i), with $v = t = 2mn$ and $w = mn$, the set of equations in (3.10) has a solution $\{\eta_1, \ldots, \eta_t\}$, not all zero, in $\mathcal{K}$ satisfying

$$
\begin{aligned}
\lceil\eta_k\rceil &< c_{3.2}\left(c_{3.3}t c_{3.10}^n n^{\frac{n-1}{2}}\right)^{\frac{mn}{t-mn}}\\
&< c_{3.11}^n n^{\frac{n-1}{2}} u^2\\
&< c_{3.12}^n n^{\frac{n+1}{2}} \text{ for } 1 \leq k \leq t.
\end{aligned}
\tag{3.12}
$$

$R(z)$ is Not a Zero Polynomial
Suppose $R(z) \equiv 0$ for any z. Then expanding the exponentials in (3.9) we have

$$\eta_1 \theta_1^k + \cdots + \eta_t \theta_t^k = 0 \text{ for } k = 0, 1, \ldots.$$

Since the determinant of the matrix $(\theta_i^j)_{i,j=1}^t$ is non-zero as $\theta_i \neq \theta_k$ for $i \neq k$, we get $\eta_i = 0$ for $1 \leq i \leq t$, a contradiction. Thus $R(z) \not\equiv 0$.

Hence there exist integers r and s^* with $r \geq n, 1 \leq s^* \leq m$ such that

$$R^{(k)}(s) = 0 \text{ for } 0 \leq k \leq r-1 \text{ and } 1 \leq s \leq m$$

and

$$R^{(r)}(s^*) \neq 0.$$

Define
$$\lambda = (\log \alpha)^{-r} R^{(r)}(s^*).$$

Thus $\lambda \neq 0$. In the rest of the proof, we find lower and upper bounds for $|\mathcal{N}(\lambda)|$ which contradict each other.

Lower Bound for $|\mathcal{N}(\lambda)|$
By (3.11), we know that
$$c_{3.7}^{r+2mu} \lambda \in \mathcal{O}_{\mathcal{K}}.$$

Hence as $r \geq n$,
$$|\mathcal{N}(\lambda)| \geq c_{3.7}^{-\nu(r+2mu)} \geq c_{3.13}^{-r}. \tag{3.13}$$

Upper Bound for $|\mathcal{N}(\lambda)|$
Note that
$$|\mathcal{N}(\lambda)| \leq \overline{|\lambda|}^{\nu-1} |\lambda|. \tag{3.14}$$

We give upper estimates for $\overline{|\lambda|}$ and $|\lambda|$.

Upper Bound for $\overline{|\lambda|}$
We have
$$\overline{|\lambda|} \leq t \max_{1 \leq k \leq t} \overline{|\eta_k|} \, \overline{|e^{s^*\theta_k}|} \, \overline{|\theta_k^r|}.$$

Now
$$\overline{|\theta_k|} = \overline{|(a+b\beta)\log\alpha|} \leq c_{3.14} u$$

and
$$\overline{|e^{s^*\theta_k}|} \leq c_{3.15}^u.$$

Hence, by (3.12),
$$\overline{|\lambda|} \leq t c_{3.12}^n n^{\frac{n+1}{2}} c_{3.15}^u (c_{3.14}u)^r \leq c_{3.16}^r r^{(2r+3)/2}. \tag{3.15}$$

Upper Bound for $|\lambda|$ Using Cauchy Integral Formula
Let
$$T(z) = r! \frac{R(z)}{(z-s^*)^r} \prod_{\substack{k=1 \\ k \neq s^*}}^{m} \left(\frac{s^*-k}{z-k} \right)^r.$$

Then observe that
$$T(s^*) = R^{(r)}(s^*) = (\log\alpha)^r \lambda.$$

Since $R^{(j)}(s^*) = 0$ for $0 \leq j < r$, $T(z)$ has a Taylor's expansion at $z = s^*$ which is valid for all $z \in \mathbb{C}$. Hence $T(z)$ is an entire function.

Applying Cauchy Integral formula for λ on $C : |z| = m(1 + \frac{r}{u})$ we get

$$\lambda = (\log \alpha)^{-r} T(s^*) = (\log \alpha)^{-r} \frac{1}{2\pi i} \int_C \frac{T(z)}{z - s^*} dz.$$

Note that for z on C, we have $s^* \leq m < |z|$. We estimate $|T(z)|$ as below. First,

$$
\begin{aligned}
|R(z)| &\leq t \max_{k,i} |\eta_k| e^{|\theta_i||z|} \\
&\leq t \max_k |\eta_k| e^{u(1+|\beta|)m(\frac{u+r}{u}) \log |\alpha|} \\
&\leq t c_{3.12}^n n^{(n+1)/2} c_{3.17}^{u+r}.
\end{aligned}
$$

Thus

$$|R(z)| \leq c_{3.18}^r r^{(r+3)/2} \tag{3.16}$$

since $r \geq n$ and $t = 2mn$. Next

$$|z - k| \geq |z| - k \geq m\left(1 + \frac{r}{u}\right) - m = \frac{mr}{u} \text{ for } 1 \leq k \leq m.$$

Hence

$$
\left| (z - s^*)^{-r} \prod_{k=1, k \neq s^*}^{m} \left(\frac{s^* - k}{z - k}\right)^r \right| \leq \left(\frac{mr}{u}\right)^{-r} \prod_{k=1}^{m} \left(\frac{u}{mr} m\right)^r
$$

$$\leq c_{3.19} \left(\frac{u}{r}\right)^{mr}.$$

We use the above inequality along with (3.16) to get

$$
\begin{aligned}
|T(z)| &\leq r! c_{3.18}^r r^{(r+3)/2} c_{3.19} \left(\frac{u}{r}\right)^{mr} \\
&\leq c_{3.20}^r r^r r^{\frac{r+3}{2}} r^{-\frac{mr}{2}} = c_{3.20}^r r^{\frac{r(3-m)+3}{2}}.
\end{aligned}
$$

Thus

$$
\begin{aligned}
|\lambda| &\leq |\log \alpha|^{-r} \frac{1}{2\pi} \left| \int_C \frac{T(z)}{z - s^*} dz \right| \\
&\leq |\log \alpha|^{-r} m\left(1 + \frac{r}{u}\right) \frac{u}{mr} c_{3.20}^r r^{\frac{r(3-m)+3}{2}}.
\end{aligned}
$$

Hence

$$|\lambda| \leq c_{3.21}^r r^{\frac{r(3-m)+3}{2}}. \tag{3.17}$$

Using (3.15) and (3.17) in (3.14), we get

$$|\mathcal{N}(\lambda)| \leq c_{3.16}^{(\nu-1)r} r^{(\nu-1)(2r+3)/2} c_{3.21}^r r^{(r(3-m)+3)/2}$$
$$\leq c_{3.22}^r r^{((\nu-1)(2r+3)+r(3-2\nu-2)+3)/2}.$$

Thus

$$|\mathcal{N}(\lambda)| \leq c_{3.22}^r r^{(3\nu-r)/2}. \tag{3.18}$$

Comparing (3.13) and (3.18), we see that

$$c_{3.13}^{-r} \leq c_{3.22}^r r^{(3\nu-r)/2}$$

or

$$r^{(r-3\nu)/2} < c_{3.23}^r.$$

Since $r \geq n$ and $c_{3.23}$ is independent of n, this inequality does not hold for n sufficiently large.

Thus we conclude that $\omega = \alpha^\beta$ is not algebraic.   □

Exercise

1. If $z \in \mathbb{C}$ is a non-rational zero of the equation

$$\sqrt{3}(1+z) = \tan(z\pi/2), \tag{3.19}$$

   then show that z is transcendental.
2. Prove that the following statement is equivalent to Theorem 3.0.1. *Let α and β be algebraic numbers such that they are $\mathbb{Q}$-linearly independent. Then, for any $t \in \mathbb{C}\backslash\{0\}$, at least one of $e^{t\alpha}$ and $e^{t\beta}$ is transcendental.*

Notes

There are still different proofs of Gelfond–Schneider theorem available now, for instance, see [1] for a proof based on the method of interpolation determinants introduced in 1992 by M. Laurent. It is known that the roots of the Eq. (3.19) with $\Re z > -1$ is simple and real. The *Amick–Fraenkel conjecture* asserts that the set $\{1, z_1, z_2, \ldots\}$ of the zeros of (3.19) is linearly independent over $\mathbb{Q}$. This is known under Schanuel's conjecture. See [2].

References

1. Yu.V. Nesterenko, *Algebraic Independence*, vol. 14 (Tata Institute of Fundamental Research Publications, Mumbai, 2008), 157 p
2. E. Shargorodsky, On the Amick-Fraenkel conjecture. Quart. J. Math. **65**, 267–278 (2014)

Chapter 4
Extensions Due to Ramachandra

Change is hard at first, messy in the middle, and gorgeous at the end

—Robin Sharma

In 1968, Ramachandra [1, 2] proved results relating to the set of complex numbers at which a given set of algebraically independent meromorphic functions assumes values in a fixed algebraic number field. These results proved to be significant in the case, to quote his own words *"(overlooked by Gelfond) where the functions concerned do not satisfy algebraic differential equations of the first order with algebraic number coefficients."* His result, besides simplifying Schneider's method, enables one to study the set of all complex numbers at which two algebraically independent meromorphic functions $f(z)$ and $g(z)$ take values which are algebraic numbers. In particular, he was able to obtain results when $(f(z), g(z)) \in \{(z, \wp(az)), (e^z, \wp(az)), (\wp_1(z), \wp_2(az))\}$ where $a \neq 0$ is an arbitrary complex number and $\wp$, $\wp_1$ and $\wp_2$ are Weierstrass elliptic functions. We refer to [2] for these results.

In this chapter we give two theorems of Ramachandra. Theorem 3.0.1 is deduced from these theorems. Further we give few other applications for instance, about the transcendence of values of Weierstrass elliptic function.

4.1 Functions Satisfying Differential Equations

Let

$$P(x_1, \ldots, x_r) = \sum_{\lambda_1, \ldots, \lambda_r} p_{\lambda_1, \ldots, \lambda_r} x_1^{\lambda_1} \ldots x_r^{\lambda_r}, \quad Q(x_1, \ldots, x_r) = \sum_{\lambda_1, \ldots, \lambda_r} q_{\lambda_1, \ldots, \lambda_r} x_1^{\lambda_1} \ldots x_r^{\lambda_r}$$

© Springer Nature Singapore Pte Ltd. 2020
S. Natarajan and R. Thangadurai, *Pillars of Transcendental Number Theory*,
https://doi.org/10.1007/978-981-15-4155-1_4

be two polynomials in $\mathbb{C}[x_1, \ldots, x_r]$. We say that the polynomial P is *majorised* by the polynomial Q and written as $P \ll Q$ if $|p_{\lambda_1,\ldots,\lambda_r}| \leq |q_{\lambda_1,\ldots,\lambda_r}|$ for every $(\lambda_1, \ldots, \lambda_r)$.

We need to study meromorphic functions when they satisfy some differential equation. The following lemma describes one such situation. For any non-zero algebraic number, we denote by $s(\alpha)$ the size of α as in Sect. 1.3. We will be using the results from Lemma 1.3.2 often without any mention.

Lemma 4.1.1 *Let $\mathcal{K}$ be a number field. Let f be a meromorphic function. For some integer $k \geq 1$, suppose f satisfies a differential equation as follows.*

$$f^{(k)}(z) = \sum_{\nu_1=0}^{n_1} \cdots \sum_{\nu_k=0}^{n_k} d_{\nu_1,\ldots,\nu_k}(f^{(0)}(z))^{\nu_1} \cdots (f^{(k-1)}(z))^{\nu_k} \text{ with } d_{\nu_1,\ldots,\nu_k} \in \mathcal{K}. \quad (4.1)$$

Suppose $f^{(0)}(z_0), \ldots, f^{(k-1)}(z_0)$ are all in $\mathcal{K}$ for some $z = z_0$. Then there exist positive integers b' and c' such that for any integer $\tau \geq 0$, we have

(i) $(b')^{\tau+1} f^{(\tau)}(z_0)$ is an algebraic integer in $\mathcal{K}$.
(ii) $s(f^{(\tau)}(z_0)) \leq (c')^{\tau+1}(\tau + 1)^{\tau}$.

Proof Let $n = 1 + \sum_{i=1}^{k} n_i$. Treating $f^{(0)}(z), \ldots, f^{(k-1)}(z)$ as variables, we see from (4.1) that $f^{(k)}(z)$ is of degree at most n. Further, $f^{(k+1)}(z)$ is of degree at most $2n$; $f^{(k+2)}(z)$ is of degree at most $3n$ and so on. Thus $f^{(k+j)}(z)$ is of degree at most $n(j + 1)$. In other words, $f^{(j)}(z)$ is of degree at most $n(j - k + 1) < n(j + 1)$ for $j \geq k$. This is obviously true for $j < k$.

Let d be the denominator of $d_{\nu_1,\ldots,\nu_k}$ and $f^{(\tau)}(z_0), 0 \leq \tau \leq k - 1$. Then we can take $b' = d^n$. This proves (i) of the lemma.

For the second part of the lemma, we take

$$Q(z) = Q(f^{(0)}(z), \ldots, f^{(k-1)}(z)) = 1 + \sum_{\nu_1=0}^{n_1} \cdots \sum_{\nu_k=0}^{n_k} d_{\nu_1,\ldots,\nu_k} + f^{(0)}(z) + \cdots + f^{(k-1)}(z).$$

We claim that for any integer $j \geq k$, we get

$$f^{(j)}(z) \ll (nQ^n(z))^{j+1}(j + 1)^j.$$

Note that by (4.1),

$$f^{(k)}(z) \ll Q^n(z),$$

i.e. $f^{(k)}(z)$ is majorised by $Q^n(z)$. Hence

$$f^{(k+1)}(z) \ll nQ^{n-1}(z)(f^{(1)}(z) + \cdots + f^{(k)}(z)) \ll nQ^{n-1}(z)(Q(z) + Q^n(z)) \ll nQ^{2n}(z)$$

from which we also get

$$f^{(k+2)}(z) \ll n(2n)Q^{2n-1}(z)Q^{n+1}(z) = 2!n^2 Q^{3n}(z).$$

Proceeding thus, we find that

$$f^{(k+j)}(z) \ll j!n^j Q^{(j+1)n}(z).$$

Therefore for $j \geq k$,

$$f^{(j)}(z) \ll (nQ^n(z))^{(j+1)}(j+1)^j, \tag{4.2}$$

as claimed. The above claim is trivially true for $j < k$. Let us choose

$$c' = n\left(1 + \sum_{\nu_1,\ldots,\nu_k} s(d_{\nu_1,\ldots,\nu_k}) + \sum_{r=0}^{k-1} s(f^{(r)}(z_0))\right)^n.$$

To complete the proof of (ii), for any integer $\tau \geq 0$, we need to estimate $s(f^{(\tau)}(z_0))$. By (4.2),

$$\left\lceil f^{(\tau)}(z_0)\right\rceil \leq (\tau+1)^\tau n^{\tau+1}\left(1 + \sum_{\nu_1,\ldots,\nu_k} \left\lceil d_{\nu_1,\ldots,\nu_k}\right\rceil + \sum_{r=0}^{k-1} \left\lceil f^{(r)}(z_0)\right\rceil\right)^{n(\tau+1)}$$

and we know by the first part (i) of the lemma that

$$d(f^{(\tau)}(z_0)) \leq d^{n(\tau+1)}.$$

Hence

$$s(f^{(\tau)}(z_0)) \leq (\tau+1)^\tau n^{\tau+1}\left(d^n + \left(1 + \sum_{\nu_1,\ldots,\nu_k} \left\lceil d_{\nu_1,\ldots,\nu_k}\right\rceil + \sum_{r=0}^{k-1} \left\lceil f^{(r)}(z_0)\right\rceil\right)^n\right)^{(\tau+1)}$$

$$\leq (\tau+1)^\tau (c')^{\tau+1}$$

proving (ii). $\qquad\qquad\qquad\qquad\qquad\qquad\qquad\qquad\qquad\qquad\Box$

In the next lemma we deal with two meromorphic functions.

Lemma 4.1.2 *Let the hypothesis of Lemma 4.1.1 be satisfied for $f = f_i$ with $k = k_i$, $i = 1, 2$. Let b'_i, c'_i be the corresponding values of b' and c'. Let $b = \max(b'_1, b'_2)$ and $c = b\max(c'_1, c'_2)$. Let ρ_1 and ρ_2 be given natural numbers. Then for any integer $j \geq 0$, we have*

(i) $b^{j+\rho_1+\rho_2} \dfrac{d^j}{dz^j}((f_1(z))^{\rho_1}(f_2(z))^{\rho_2})\,|_{z=z_0}$ *is an algebraic integer.*

(ii) $\quad s\left(\dfrac{d^j}{dz^j}((f_1(z))^{\rho_1}(f_2(z))^{\rho_2})\mid_{z=z_0}\right) \leq (\rho_1 + \rho_2)^j (j+1)^j c^{j+\rho_1+\rho_2}.$

Proof We have

$$\frac{d^j}{dz^j}((f_1(z))^{\rho_1}(f_2(z))^{\rho_2}) = \sum_{\nu=0}^{j} \binom{j}{\nu} \frac{d^\nu}{dz^\nu}((f_1(z))^{\rho_1})\frac{d^{j-\nu}}{dz^{j-\nu}}((f_2(z))^{\rho_2}).$$

Now

$$\frac{d^\nu}{dz^\nu}((f_1(z))^{\rho_1}) = \sum_{\mu_1+\cdots+\mu_{\rho_1}=\nu} f_1^{(\mu_1)}(z)\cdots f_1^{(\mu_{\rho_1})}(z).$$

Hence by Lemma 4.1.1

$$b^{\nu+\rho_1}\frac{d^\nu}{dz^\nu}((f_1(z))^{\rho_1})\mid_{z=z_0}$$

is an algebraic integer. Similarly,

$$b^{j-\nu+\rho_2}\frac{d^{j-\nu}}{dz^{j-\nu}}((f_2(z))^{\rho_2})\mid_{z=z_0}$$

is an algebraic integer. Thus we get (i).

(ii) Since $b^{j+1}f_i^{(j)}(z_0)$ is an algebraic integer we get that

$$s\left(b^{\rho_1+\nu}\frac{d^\nu}{dz^\nu}(f_1^{\rho_1})\mid_{z=z_0}\right) \leq \sum_{\mu_1+\cdots+\mu_{\rho_1}=\nu} s(b^{\mu_1+1}f_1^{(\mu_1)}\cdots b^{\mu_{\rho_1}+1}f_1^{(\mu_{\rho_1})}\mid_{z=z_0}).$$

$$\leq \sum_{\mu_1+\cdots+\mu_{\rho_1}=\nu} s(b^{\mu_1+1}f_1^{(\mu_1)})\mid_{z=z_0}\cdots s(b^{\mu_{\rho_1}+1}f_1^{(\mu_{\rho_1})})\mid_{z=z_0}.$$

Note that by Lemma 4.1.1(ii), the right-hand side of the above inequality is bounded by

$$b^{\rho_1+\nu}\left(\sum_{\mu_1+\cdots+\mu_{\rho_1}=\nu} 1\right)(c_1')^{\rho_1+\nu}\prod_{k=1}^{\rho_1}(\mu_k+1)^{\mu_k}.$$

Thus

$$s\left(\frac{d^\nu}{dz^\nu}(f_1^{\rho_1})\mid_{z=z_0}\right) \leq \rho_1^\nu(bc_1')^{\rho_1+\nu}(\nu+1)^\nu \qquad (4.3)$$

since $\displaystyle\sum_{\mu_1+\cdots+\mu_{\rho_1}=\nu} 1$ is the value of $(x_1+\cdots+x_{\rho_1})^\nu$ when $x_1=\cdots=x_{\rho_1}=1$ and

$$\prod_{k=1}^{\rho_1}(\mu_k+1)^{\mu_k} \le \left(\sum_{k=1}^{\rho_1}\mu_k+1\right)^{\sum_{k=1}^{\rho_1}\mu_k}.$$

Similar inequality as (4.3) holds for $f_2^{\rho_2}$ with ρ_1, c_1' replaced by ρ_2, c_2'. Hence

$$s\left(\frac{d^j}{dz^j}(f_1^{\rho_1}f_2^{\rho_2})\mid_{z=z_0}\right) \le \sum_{\nu=0}^{j}\binom{j}{\nu}\rho_1^\nu\rho_2^{j-\nu+1}(j-\nu+1)^{j-\nu}(\nu+1)^\nu(c')^{j+\rho_1+\rho_2}$$

$$\le (\rho_1+\rho_2)^j(j+1)^j c^{j+\rho_1+\rho_2}$$

proving (ii). □

4.2 First Extension

Let $F_1, \ldots, F_s$ be algebraically independent entire functions and let $\{(a_\mu, n_\mu)\}$ be a sequence of pairs with $a_\mu \in \mathbb{C}$ and $n_\mu \in \mathbb{N}$ with $\{n_\mu\}$ a non-decreasing sequence. We make the following hypotheses.

H1. Let $F_1, \ldots, F_s$ be entire functions of finite order $\le \rho$.
H2. Let $N(Q) = |\{a_\mu : n_\mu \le Q\}|$. Assume that $N(Q)$ is finite for any $Q \ge 1$.
H3. Let $D(Q) = \max_{n_\mu \le Q}(|a_\mu|)$.
H4. Assume that $\liminf \frac{\log N(Q)}{\log D(Q)} > \rho$.
H5. Assume that $F_t(a_\mu)$ are all algebraic. Let $\nu(Q)$ be the degree of the field $\mathbb{Q}(F_t(a_\mu))$, $1 \le t \le s$ and $n_\mu \le Q$.
H6. Let $M^{(t)}(R) = 1 + \max_{|z|=R}|F_t(z)|$, $1 \le t \le s$.
H7. Put $M_1^{(t)}(Q) = 1 + \max_{n_\mu \le Q}\{s(F_t(a_\mu))\}$, $1 \le t \le s$.

Let $f(q)$ be a positive and increasing function. Let $s \ge 1$ and $r_1, \ldots, r_s$ be positive integers. We say that

$$r_1 \cdots r_s \sim f(q)$$

if given $\epsilon > 0$, there exists $q_0(\epsilon)$ such that for $q \ge q_0(\epsilon)$ we have

$$(1-\epsilon)f(q) < r_1 \cdots r_s < (1+\epsilon)f(q).$$

For instance, suppose $f(q) = q$ and $s = 2$, then choosing $r_1 = r_2 = [q^{1/2}]$, we see that $r_1 r_2 \sim f(q)$.

Theorem 4.2.1 *Let $F_1, \ldots, F_s$ be algebraically independent entire functions and let $\{(a_\mu, n_\mu)\}$ be a sequence of pairs with $a_\mu \in \mathbb{C}$ and $n_\mu \in \mathbb{N}$ with $\{n_\mu\}$ a non-decreasing*

*sequence. Suppose the hypotheses $H1 - H7$ are satisfied. Let $q \geq 1$ and $r_1, \ldots, r_s$
be natural numbers such that*

$$r_1 \cdots r_s \sim \nu(q)(\nu(q) + 1)N(q).$$

Then there exists $Q > q$ such that for any positive number R, we have

$$\left(\prod_{t=1}^{s} (M_1^{(t)}(Q))^{r_t} \right)^{8\nu(Q)} \prod_{t=1}^{s} (M^{(t)}(R))^{r_t} \left(\frac{8D(Q)}{R} \right)^{N(Q-1)} \geq 1. \qquad (4.4)$$

Proof Note that (4.4) is trivially true if $R < 2D(Q)$. So we assume from now on
that $R \geq 2D(Q)$.

Auxiliary Polynomial
Take the entire function

$$R(z) = \sum_{k_1=0}^{r_1-1} \cdots \sum_{k_s=0}^{r_s-1} c_{k_1,\ldots,k_s} (F_1(z))^{k_1} \cdots (F_s(z))^{k_s}$$

where $c_{k_1,\ldots,k_s}$ are rational integers to be chosen soon. Consider

$$R(a_\mu) = 0 \text{ for all } a_\mu \text{ with } n_\mu \leq q. \qquad (4.5)$$

This is a system of $N(q)$ linear equations in $q_1 = r_1 \cdots r_s$ unknowns $c_{k_1,\ldots,k_s}$. Let d
be a denominator of the coefficients of this linear system. Then

$$d \leq (M_1^{(1)}(q))^{r_1} \cdots (M_1^{(s)}(q))^{r_s} =: J_1(q), \text{ say.} \qquad (4.6)$$

Upper Bound for $|c_{k_1,\ldots,k_s}|$
The size of the algebraic integer coefficients so obtained is bounded by $J_1(q)^2$. Hence
by Lemma 3.1.3 (ii), we get rational integers $c_{k_1,\ldots,k_s}$, not all zero with

$$|c_{k_1,\ldots,k_s}| < 1 + \left(2r_1 \cdots r_s J_1(q)^2 \right)^{\frac{N(q)\nu(q)(\nu(q)+1)}{2q_1 - N(q)\nu(q)(\nu(q)+1)}}$$

satisfying (4.5). Since by assumption $q_1 \sim \nu(q)(\nu(q) + 1)N(q)$, we find that given
any $\epsilon > 0$ there exists $q_0(\epsilon)$ such that for $q > q_0(\epsilon)$, the exponent above satisfies

$$\frac{N(q)\nu(q)(\nu(q) + 1)}{2q_1 - N(q)\nu(q)(\nu(q) + 1)} < \frac{1}{1 - \epsilon}.$$

Using the inequality $2r_1 \cdots r_s \leq 2^{r_1} \cdots 2^{r_s} \leq J_1(q)$, we get therefore that

$$|c_{k_1,\ldots,k_s}| < 1 + (J_1(q))^{\frac{3}{1-\epsilon}} \text{ for } q \geq q_0(\epsilon).$$

$R(a_\mu) \neq 0$ for Some a_μ

By the algebraic independence of $F_1, \ldots, F_s$, we see that

$$R(z) \not\equiv 0.$$

Further it is an entire function of order $\leq \rho$. Suppose $R(a_\mu) = 0$ for all a_μ. Then

$$\liminf \frac{\log N(Q)}{\log D(Q)} \leq \rho$$

which contradicts $H4$. Hence there exist points a_j in $\{a_\mu\}$ for which $R(a_j) \neq 0$. Choose such an a_j with least possible n_j say, $n_j = Q$. Then $Q > q$,

$$\gamma = R(a_j) \neq 0$$

and $R(a_\ell) = 0$ for all ℓ with $n_\ell < Q$.

Lower Bound for $|\mathcal{N}(\gamma)|$

By (4.6), there exists a natural number $d \leq J_1(Q)$ such that $d\gamma$ is an algebraic integer of degree $\leq \nu(Q)$ and hence

$$|\mathcal{N}(\gamma)| \geq J_1(Q)^{-\nu(Q)}. \tag{4.7}$$

Upper Bound for $|\mathcal{N}(\gamma)|$ Using Cauchy Integral Formula

Note that

$$s(\gamma) \leq r_1 \cdots r_s(1 + J_1(q)^{\frac{3}{1-\epsilon}})J_1(Q) \leq (J_1(Q))^{2+\frac{3}{1-\epsilon}}. \tag{4.8}$$

Integrating on $C : |z| = R_0$ with $R_0 \geq 2D(Q)$, we get

$$\gamma = \frac{1}{2\pi i} \int_C R(z) \prod_{n_\ell < Q} \left(\frac{a_j - a_\ell}{z - a_\ell}\right) \frac{dz}{z - a_j}.$$

We bound the integrand as follows.

(i) By putting $J_2(R_0) := \prod_{t=1}^{s} (M^{(t)}(R_0))^{r_t}$, we get

$$|R(z)| \leq r_1 \cdots r_s \left(1 + J_1(q)^{\frac{3}{1-\epsilon}}\right) J_2(R_0) \leq J_1(Q)^{1+\frac{3}{1-\epsilon}} J_2(R_0).$$

(ii) $\left| \prod_{n_\ell < Q} \frac{a_j - a_\ell}{z - a_\ell} \right| \leq \left(\frac{2D(Q)}{R_0 - D(Q)}\right)^{N(Q-1)} \leq \left(\frac{4D(Q)}{R_0}\right)^{N(Q-1)}$ since $D(Q) \leq R_0/2$.

(iii) $\dfrac{1}{|z - a_j|} \leq \dfrac{1}{R_0 - D(Q)} \leq \dfrac{2}{R_0}.$

Using (i)–(iii) in the integral expression for γ we get

$$|\gamma| \le J_1(Q)^{1+\frac{3}{1-\epsilon}} J_2(R_0) \left(\frac{8D(Q)}{R_0}\right)^{N(Q-1)}.$$

Together with (4.8), we get

$$|\mathcal{N}(\gamma)| \le J_1(Q)^{(\nu(Q)-1)(2+\frac{3}{1-\epsilon})+1+\frac{3}{1-\epsilon}} J_2(R_0) \left(\frac{8D(Q)}{R_0}\right)^{N(Q-1)}.$$

The exponent of $J_1(Q)$ is bounded by $\nu(Q)\left(3+\frac{3}{1-\epsilon}\right) = 7\nu(Q)$ by choosing $\epsilon = 1/4$. Thus

$$|\mathcal{N}(\gamma)| \le J_1(Q)^{7\nu(Q)} J_2(R_0) \left(\frac{8D(Q)}{R_0}\right)^{N(Q-1)}. \tag{4.9}$$

Final Step
Combining (4.9) with (4.7), the assertion of the theorem follows for any $R_0 \ge 2D(Q)$. This completes the proof of the theorem. $\qquad\square$

4.3   Theorem 4.2.1 Implies Theorem 4.1.1

We assume that α^β is algebraic with α, β algebraic, $\log \alpha \ne 0$ and β irrational. Let $c_{4.1}, c_{4.2}, \ldots$ denote numbers depending only on α and β. In Theorem 4.2.1 take

$$s = 2;\ F_1(z) = z,\ F_2(z) = e^{z\log\alpha};\ a_\mu = \ell + m\beta,\ \ell, m \in \mathbb{N};\ n_\mu = \max(\ell, m).$$

Since $\log \alpha \ne 0$, by Lemma 1.1.1, $F_1(z)$ and $F_2(z)$ are algebraically independent. Note that

$$\rho = 1;\ M^{(1)}(R) = 1 + R \le c_{4.1}R;\ M^{(2)}(R) \le 1 + c_{4.2}^R \le c_{4.3}^R;$$

$$N(Q) \le Q^2;\ c_{4.4} < D(Q)/Q < c_{4.5}.$$

Thus

$$\lim_{Q\to\infty} \frac{\log N(Q)}{\log D(Q)} = 2 > \rho.$$

Let $\nu(Q) = \nu$. Note that ν is the degree of $\mathbb{Q}(\alpha, \beta, \alpha^\beta)$ over $\mathbb{Q}$ which is independent of Q. Further

$$M_1^{(1)}(Q) \ge c_{4.6}Q \text{ and } M_1^{(2)}(Q) \ge c_{4.7}^Q.$$

We need to satisfy

$$r_1 r_2 \sim \nu(\nu + 1)q^2$$

which is possible by taking

$$r_1 = [\nu(\nu + 1)q^{3/2}] < \nu(\nu + 1)Q^{3/2}$$

and

$$r_2 = [q^{1/2}] < Q^{1/2}.$$

Now take $R = 16c_{4.5}Q$. Then the inequality in Theorem 4.2.1 implies that

$$((c_{4.6}Q)^{\nu(\nu+1)Q^{3/2}} c_{4.7}^{Q^{3/2}})^{8\nu}(1 + 16c_{4.5}Q)^{\nu(\nu+1)Q^{3/2}}(1 + c_{4.3}^{16c_{4.5}Q})^{Q^{1/2}} 2^{-Q^2} \geq 1.$$

This is not possible for sufficiently large Q. $\qquad\square$

4.4 Another Consequence of Theorem 4.2.1

Theorem 4.4.1 *At least one of* $2^\pi, 2^{\pi^2}, 2^{\pi^3}$ *is transcendental.*

Proof Assume that $2^\pi, 2^{\pi^2}, 2^{\pi^3}$ are all algebraic. Let $c_{4.8}, c_{4.9}, \ldots$ denote absolute constants. In Theorem 4.2.1 take

$$s = 2; \ F_1(z) = 2^z, \ F_2(z) = 2^{\pi z}; \ a_\mu = k + \ell\pi + m\pi^2, k, \ell, m \in \mathbb{N}; \ n_\mu = \max(k, \ell, m).$$

By Lemma 1.1.2, $F_1(z)$ and $F_2(z)$ are algebraically independent. Note that

$$\rho = 1; \ M^{(1)}(R) \leq c_{4.8}^R; \ M^{(2)}(R) \leq c_{4.9}^R; \ N(Q) \leq Q^3; \ c_{4.10} < D(Q)/Q < c_{4.11}.$$

Thus

$$\lim_{Q \to \infty} \frac{\log N(Q)}{\log D(Q)} = 3 > \rho.$$

Let $\nu(Q) = \nu$. Note that ν is the degree of $\mathbb{Q}(2^\pi, 2^{\pi^2}, 2^{\pi^3})$ over $\mathbb{Q}$ which is independent of Q. Further $M_1^{(1)}(Q) \leq c_{4.12}^Q$ and $M_1^{(2)}(Q) \leq c_{4.13}^Q$. We need to satisfy

$$r_1 r_2 \sim \nu(\nu + 1)q^3$$

which is possible by taking

$$r_1 = r_2 = [(\nu(\nu + 1)q^3)^{1/2}] < (\nu(\nu + 1)Q^3)^{1/2}.$$

Now take $R = 16c_{4.11}Q$. Then the inequality in Theorem 4.2.1 implies that

$$((c_{4.12}c_{4.13})^{\nu(\nu+1)^{1/2}Q^{5/2}})^{8\nu}(c_{4.8}c_{4.9})^{16c_{4.11}Q^{5/2}(\nu(\nu+1))^{1/2}}2^{-Q^3} \geq 1.$$

This is not possible for sufficiently large Q.

4.5 Second Extension

Theorem 4.5.1 *Let $i = 1, 2$. Suppose each f_i is a meromorphic function which is quotients of entire functions of order $\leq \mu$ and satisfying a differential equation of the form*

$$f_i^{(k_i)}(z) = \sum_{\nu_1=0}^{n_1} \cdots \sum_{\nu_{k_i}=0}^{n_{k_i}} d_{\nu_1,\ldots,\nu_{k_i}} (f_i^{(0)}(z))^{\nu_1} \cdots (f_i^{(k_i-1)}(z))^{\nu_{k_i}-1}$$

where all the coefficients $d_{\nu_1,\ldots,\nu_{k_i}}$ belong to a number field $\mathcal{K}$ of degree ν. Suppose there exists an infinite sequence of distinct points $\{z_\lambda\}$, $\lambda \geq 0$ such that

$$f_i^{(\ell)}(z_\lambda) \in \mathcal{K} \text{ for } 0 \leq \ell \leq k_i - 1.$$

Then f_1 and f_2 are algebraically dependent.

Proof The proof is similar to the proofs of Theorems 3.0.1 and 4.2.1.

Auxiliary Polynomial

Suppose f_1 and f_2 are algebraically independent. Let us consider a function

$$\Phi(z) = \sum_{\rho_1=0}^{r} \sum_{\rho_2=0}^{r} C_{\rho_1,\rho_2}(f_1(z))^{\rho_1}(f_2(z))^{\rho_2}$$

with

$$\Phi^{(s)}(z_\lambda) = 0 \text{ for } 0 \leq s \leq t - 1, 0 \leq \lambda \leq m - 1.$$

Here m and t are free parameters and $r = [mt\nu(\nu + 1)]$.

Coefficients C_{ρ_1,ρ_2} are in $\mathcal{K}$ and Not All Zero

Note that there are mt equations in $(r + 1)^2$ unknowns. By Lemma 4.1.1 corresponding to each z_λ, there exists positive integers b_λ and c_λ such that

$$b_\lambda^{t+2r} \Phi^{(s)}(z_\lambda) = 0 \text{ for } 0 \leq s \leq t - 1, 0 \leq \lambda \leq m - 1$$

have algebraic integer coefficients, and the size of the coefficients is bounded by

$$(2r)^t (t+1)^t (b_\lambda c_\lambda)^{t+2r} \le t^{\delta_1 t} \gamma_1^t$$

as seen in (4.3). Here and in the sequel, $\gamma_1, \gamma_2, \ldots$ denote numbers depending on m while $\delta_1, \delta_2, \ldots$ denote numbers depending on ν but independent of m and t. A priori, these numbers depend on $\mathcal{K}$. By Lemma 3.1.3, there exist rational integers C_{λ_1,λ_2}, not all zero, satisfying the given system of linear equations and such that

$$|C_{\lambda_1,\lambda_2}| \le 1 + (t^{\delta_1 t} \gamma_1^t)^{\frac{m t \nu(\nu+1)}{(r+1)^2 - m t \nu(\nu+1)}} \le t^{\delta_2 t} \gamma_2^t. \tag{4.10}$$

By construction above, $\Phi(z) \not\equiv 0$. So it has zeros of finite order at z_λ, $0 \le \lambda \le m-1$. Hence there exists $j \ge t$ with

$$\Phi^{(\tau)}(z_\lambda) = 0 \text{ for } 0 \le \tau \le j-1; 0 \le \lambda \le m-1$$

but

$$\Phi^{(j)}(z_{\lambda_0}) \ne 0 \text{ for some } \lambda_0 \text{ with } 0 \le \lambda_0 \le m-1.$$

We shall find lower and upper bounds for $|\Phi^{(j)}(z_{\lambda_0})|$.

Lower Bound for $|\Phi^{(j)}(z_{\lambda_0})|$
We know

$$b_{\lambda_0}^{j+2r} \Phi^{(j)}(z_{\lambda_0})$$

is a non-zero algebraic integer. Hence

$$|\mathcal{N}(b_{\lambda_0}^{j+2r} \Phi^{(j)}(z_{\lambda_0}))| \ge 1. \tag{4.11}$$

On the other hand, by (4.10) and Lemma 4.1.2 (ii), there exists a positive integer c_{λ_0} such that

$$s\left(\Phi^{(j)}(z_{\lambda_0})\right) \le (r+1)^2 \gamma_2^t t^{\delta_2 t} (2r)^j (j+1)^j c_{\lambda_0}^{j+2r} \le j^{\delta_3 j} \gamma_3^j.$$

Using this in (4.11), we get

$$|\Phi^{(j)}(z_{\lambda_0})| > j^{-\delta_4 \nu j} \gamma_4^j = j^{-\delta_5 j} \gamma_4^j. \tag{4.12}$$

Upper Bound for $|\Phi^{(j)}(z_{\lambda_0})|$ Using Cauchy Integral Formula
Write

$$f_i = h_i / g_i, i = 1, 2$$

with h_i, g_i entire functions of order $\le \mu$. Consider $G = (g_1 g_2)^r$. By the hypothesis, g_1 and g_2 do not vanish at $z_0, \ldots, z_{m-1}$. Hence

$$\left| \frac{1}{G(z_{\lambda_0})} \right| \le \gamma_5^{2r} \le \gamma_6^t.$$

Consider the integral

$$I = \frac{1}{2\pi i} \int_T \frac{\Phi(z)G(z)dz}{(z-z_{\lambda_0}) \prod_{\lambda=0, \lambda \neq \lambda_0}^{m-1} (z-z_\lambda)^j}$$

where $T : |z| = j^\delta$ with δ to be chosen later. Note that since $j \geq t$ by taking t sufficiently large, we may assume that $\frac{1}{2}T =: |z| = \frac{1}{2}j^\delta$ contains $z_0, \ldots, z_{m-1}$. By Cauchy Integral formula,

$$I = \frac{1}{j!} \frac{G(z_{\lambda_0})}{\prod_{\lambda=0, \lambda \neq \lambda_0}^{m-1} (z_{\lambda_0} - z_\lambda)^j} \Phi^{(j)}(z_{\lambda_0}).$$

Hence

$$\Phi^{(j)}(z_{\lambda_0}) = \frac{j! \prod_{\lambda=0, \lambda \neq \lambda_0}^{m-1} (z_{\lambda_0} - z_\lambda)^j}{G(z_{\lambda_0})} \cdot I.$$

Note that

$$\left| \prod_{\lambda=0, \lambda \neq \lambda_0}^{m-1} (z_{\lambda_0} - z_\lambda)^j \right| < \gamma_7^j.$$

We now use the fact that f_1, f_2 are meromorphic functions of order $\leq \mu$. Hence for any given $\epsilon > 0$, there exist $j_0(\epsilon)$ such that for $j \geq j_0(\epsilon)$ we have

$$\max_{|z|=j^\delta} |\Phi(z)G(z)| < (r+1)^2 \gamma_2^t t^{\delta_2 t} e^{2rj^{\delta(\mu+\epsilon)}}$$

by using also (4.10). Choose $\delta = 1/(2\mu + 1)$. Then for $\epsilon < 1/2, \delta(\mu + \epsilon) < 1/2$. Hence

$$\max_{|z|=j^\delta} |\Phi(z)G(z)| < \gamma_8^j j^{\delta_6 j}$$

and

$$\min_{|z|=j^\delta} \left| (z-z_{\lambda_0}) \prod_{\lambda=0}^{m-1} |(z-z_\lambda)^j| \geq 2^{-(mj+1)} j^{\delta(mj+1)}.$$

Hence by all the above estimates, we have

$$|\Phi^{(j)}(z_{\lambda_0})| < j! \gamma_6^t \gamma_7^j \gamma_8^j j^{\delta_6 j} 2^{mj+1} j^{-\delta(mj+1)} j^\delta < \gamma_9^j j^{(\delta_6+1-m\delta)j}$$

for $j \geq j_0$ and hence for $t \geq t_0$ where t_0 is sufficiently large. Comparing this upper bound for $|\Phi^{(j)}(z_{\lambda_0})|$ with the lower bound in (4.12), we get

$$j^{\delta_5 + \delta_6 + 1 - m\delta} > \frac{\gamma_4}{\gamma_9}.$$

This is not valid for large m. This completes the proof of Theorem 4.5.1.   □

4.6 Some Consequences of Theorem 4.5.1

Derivation of Gelfond–Schneider Theorem
In Theorem 4.5.1 take

$$f_1(z) = e^z, \ f_2(z) = e^{\beta z} \text{ with } \beta \text{ algebraic irrational}; \ z_\lambda = \lambda \log \alpha, \ \lambda = 0, 1, \dots$$

with α algebraic having $\log \alpha \neq 0$. If α^β were algebraic, then by Theorem 4.5.1, f_1 and f_2 are algebraically dependent which contradicts Lemma 1.1.2.   □

Another Consequence of Theorem 4.5.1
For basic definition and properties of functions involved for the rest of the discussion, we refer to Apostol [3].

Theorem 4.6.1 *Let $j(z)$ be the modular invariant function associated with $\wp(z)$. If τ is algebraic complex number with its imaginary part positive but not imaginary quadratic, then $j(\tau)$ is transcendental.*

It is well known that $j(\tau)$ is algebraic if τ is imaginary quadratic (see [4]). We first prove a lemma on Weierstrass $\wp$-function which itself is interesting.

Lemma 4.6.2 *Let g_2, g_3 be invariants of $\wp(z)$ and g_2^*, g_3^* be the invariants of $\wp^*(z)$. Let $\wp(z)$ and $\wp^*(\beta z)$ be algebraically independent. Then for any $\alpha \in \mathbb{C}$ which is not a pole of $\wp(z)$ and $\wp^*(\beta z)$, one at least of the seven numbers*

$$g_2, g_3, g_2^*, g_3^*, \beta, \wp(\alpha), \wp^*(\beta \alpha) \tag{4.13}$$

is transcendental.

Proof Suppose all the seven numbers in (4.13) are algebraic. We take $z_\lambda = \lambda \alpha$, $\lambda = 0, 1, 2, \dots$ except those non-negative integers for which $\lambda \alpha$ is a pole of either $\wp(z)$ or $\wp^*(\beta z)$. We know

$$\wp'^2 = 4\wp^3 - g_2\wp - g_3; \ \wp'' = 6\wp^2 - g_2/2.$$

Hence $\wp^{(i)}(\alpha), i \geq 0$ are all algebraic. Similarly $(\wp^*)^{(i)}(\beta\alpha), i \geq 0$ are all algebraic. We also know

$$\wp(z + z^*) = -\wp(z) - \wp(z^*) + \frac{1}{4}\left(\frac{\wp'(z) - \wp'(z^*)}{\wp(z) - \wp(z^*)}\right)$$

for $z \not\equiv z^* \pmod{(\omega_1, \omega_2)}$ where ω_1 and ω_2 are periods of $\wp$ and

$$\wp(2z) = -2\wp(z) + \frac{\wp''}{4\wp'}.$$

Let $\lambda = 1, 2, \ldots$. If $\lambda\alpha$ is not a pole of $\wp(z)$, then $\wp(\lambda\alpha)$ is expressible in terms of $\wp^{(i)}(\alpha)$ by applying L'Hospital's rule if necessary. Thus $\wp(\lambda\alpha)$ are all algebraic. Similarly $\wp^*(\lambda\beta\alpha)$ is expressible in terms of $(\wp^*)^{(i)}(\beta\alpha)$ and so $\wp^*(\lambda\beta\alpha)$ are all algebraic. By assumption, all these lie in the field generated by the seven algebraic numbers in (4.13) and $\wp'(\alpha), (\wp^*)'(\beta\alpha)$. Hence $\wp(z)$ and $\wp^*(\beta z)$ are algebraically dependent by Theorem 4.5.1, a contradiction to Lemma 1.1.3.

Proof of Theorem 4.6.1 Assume that $j(\tau)$ is algebraic. Write $\tau = \omega_2/\omega_1$. We have

$$j(\tau) = g_2^3/(g_2^3 - 27g_3^2).$$

If $j(\tau) = 0$, then $g_2 = 0$ and ω_1, ω_2 may be normalised to give $g_3 = 1$. If $j(\tau) \neq 0$, then $g_2 \neq 0$ and normalise g_2 to be 1 and then g_3 has to be algebraic since $j(\tau)$ is algebraic. Thus

$$\tau, g_2, g_3$$

are all algebraic. Since

$$4(\wp'(z))^2 = (\wp(z) - \wp(\omega_1/2))(\wp(z) - \wp(\omega_2/2))(\wp(z) - \wp((\omega_1 + \omega_2)/2)),$$

the numbers $\wp(\omega_1/2), \wp(\omega_2/2)$ and $\wp((\omega_1 + \omega_2)/2)$ are all algebraic.

Next set $\omega_1^* = \omega_1\tau, \omega_2^* = \omega_2\tau$. Let $\wp^*(z)$ be the corresponding Weierstrass elliptic function. Then $g_2^*, g_3^*, \wp^*(\omega_1^*/2), \wp^*(\omega_2^*/2), \wp^*((\omega_1^* + \omega_2^*)/2)$ are all algebraic. Further since

$$\wp^*(\omega_2/2) = \wp^*\left(\frac{\omega_2^*}{2}\tau^{-1}\right) = \tau^2\wp\left(\frac{\omega_2^*}{2}\right),$$

we find that $\wp^*(\omega_2/2)$ is algebraic. Thus $g_2, g_3, g_2^*, g_3^*, 1, \wp(\omega_2/2), \wp^*(\omega_2/2)$ are algebraic and hence $\wp(z), \wp^*(z)$ are algebraically dependent by Lemma 4.6.2. So their periods are commensurable by Lemma 1.1.3, i.e. there exist non-zero integers n_1, n_2 such that

$$n_1\omega_1\tau = a\omega_1 + b\omega_2; n_2\omega_2\tau = c\omega_1 + d\omega_2$$

with a, b, c, d integers and b, d non-zero. Hence

$$\tau \frac{n_2}{n_1} = \frac{c + d\tau}{a + b\tau}$$

leading to a quadratic equation for τ, a contradiction. $\square$

Exercise

1. Let $\alpha, \beta, \gamma, \delta$ be algebraic numbers with $\log \alpha, \log \beta, \log \delta$ are $\mathbb{Q}$-linearly independent and $\log \gamma / \log \delta \notin \mathbb{Q}$. Show that at least one of

$$\alpha^{\log \gamma / \log \delta} \text{ and } \beta^{\log \gamma / \log \delta}$$

is transcendental.

Notes

By using the proof of Theorem 4.4.1, one can show the following result, known as *six exponentials theorem.*

Let $\{a_1, a_2\}$ and $\{b_1, b_2, b_3\}$ be two sets of $\mathbb{Q}$−linearly independent complex numbers. Then at least one of the six numbers

$$\exp(a_i b_j), 1 \le i \le 2, 1 \le j \le 3,$$

is transcendental.

The corresponding result for four exponentials *still* remains *unsolved.* A particular case of the four exponential result was conjectured in 1944 by Alaoglu and Erdős [5]:

For any distinct prime numbers p and q, if p^x and q^x are integers for some real number x, then x must be an integer.

This case also remains *unsolved.* For related details, see [6]. A good source for various transcendence and algebraic independence results on the values of the exponential function is the book of Waldschmidt [7].

A simultaneous approximation measure for the three numbers 2^π, 2^{π^2} and 2^{π^3} was obtained by Shorey in 1974; see [8]. This was generalised to 2^{π^k} for $k = 1, 2, \ldots$ by Srinivasan; see [9, 10].

In 1962, Lang generalised Schneider's method which is similar to Theorem 4.5.1 as follows.

Let $\mathcal{K}$ be a number field. Let $f_1, f_2, \ldots, f_n$ be meromorphic functions of order $\le \rho$ such that at least two of these functions are algebraically independent. Suppose that the derivative $D = d/dz$ as a map takes the ring $\mathcal{K}[f_1, f_2, \ldots, f_n]$ into itself. If $\alpha_1, \alpha_2, \ldots, \alpha_m$ are distinct elements of $\mathbb{C}$ not among the poles of f_i such that $f_i(\alpha_j) \in \mathcal{K}$ for all $1 \le i \le n$ and $1 \le j \le m$, then $m \le 4\rho[\mathcal{K} : \mathbb{Q}]$.

For its proof and some consequences, we refer to [4]. The above result was further generalised by Bombieri; see [11].

References

1. K. Ramachandra, Contributions to the theory of transcendental numbers I. Acta Arithmetica **14**, 65–72 (1968); II **14**, 73–88 (1968)
2. K. Ramachandra, *Lectures on Transcendental Numbers* (The Ramanujan Institute, University of Madras, Chennai, 1969)
3. T.M. Apostol, *Modular Functions and Dirichlet Series in Number Theory*, 2nd edn. Graduate Texts in Mathematics, vol. 41 (Springer, New York, 1990)
4. M. Ram Murty, P. Rath, *Transcendental Numbers* (Springer, Berlin, 2014), 217 pp
5. L. Alaoglu, P. Erdős, On highly composite and similar numbers. Trans. Amer. Math. Soc. **56**, 448–469 (1944)
6. K. Senthil Kumar, R. Thangadurai, V. Kumar, On a problem of Alaoglu and Erdős. Resonance **23**(7), 749–758 (2018)
7. M. Waldschmidt, *Nombres Transcendants* (Springer, Berlin, 1974)
8. T.N. Shorey, On the sum $\sum_{k=1}^{3} \left| 2^{\pi^k} - \alpha_k \right|$, α_k algebraic numbers. J. Number Theory **6**, 248–260 (1974)
9. S. Srinivasan, On algebraic approximation to 2^{π^k} ($k = 1, 2, 3, \ldots$), I. Indian J. Pure Appl. Math. **5**, 513–523 (1974)
10. S. Srinivasan, On algebraic approximation to 2^{π^k} ($k = 1, 2, 3, \ldots$), II. J. Indian Math. Soc. (N.S.) **43** (1979); (1–4), 53–60 (1980)
11. Yu.V. Nesterenko, *Algebraic Independence*, vol. 14 (Tata Institute of Fundamental Research Publications, Mumbai, 2008), 157 pp

Chapter 5
Diophantine Approximation and Transcendence

Come friends, it's not too late to seek a newer world

—Tennyson

Diophantine approximation deals with the solubility of inequalities in integers. Dirichlet obtained one of the first type of such result in 1842 based on *pigeon-hole* principle. He showed that when α is irrational, there exist infinitely many rationals $p/q(q > 0)$ such that

$$\left| \alpha - \frac{p}{q} \right| < \frac{1}{q^2}.$$

In 1844, Liouville proved that for any algebraic number α of degree $n \geq 2$, there exists a computable number $c(\alpha) > 0$ such that

$$\left| \alpha - \frac{p}{q} \right| > \frac{c(\alpha)}{q^\kappa}$$

with $\kappa = n$. This led him to construct first examples of transcendental numbers. When α is a quadratic irrational, then $\kappa = 2$, and this cannot be improved by the first inequality of Dirichlet above. Thus in Liouville's result $\kappa = 2 + \epsilon, \epsilon > 0$ is essentially the best exponent to be expected.

In 1909, the Norwegian mathematician Thue used the approximation techniques in an ingenious way to make a major advancement to solve certain equations. Thus he could show that the equation

$$F(x, y) = m$$

© Springer Nature Singapore Pte Ltd. 2020
S. Natarajan and R. Thangadurai, *Pillars of Transcendental Number Theory*,
https://doi.org/10.1007/978-981-15-4155-1_5

where F is an irreducible binary form with integral coefficients of deg ≥ 3, possesses only *finitely* many solutions in integers x and y. For this, he realised that the exponent κ should be lowered. He showed that for any given $\epsilon > 0$, one can take $\kappa = n/2 + 1 + \epsilon$. Sections 6.1 and 6.2 describe the results of Dirichlet, Liouville and Thue.

Siegel improved Thue's result, by showing that

$$\kappa = \min_{1 \leq s \leq n-1, s \in \mathbb{Z}} \left(\frac{n}{s+1} + s \right) + \epsilon. \tag{5.1}$$

In particular, κ can be taken as $2\sqrt{n} + \epsilon$. The next improvement was by Dyson in 1947 with $\kappa = \sqrt{2n} + \epsilon$. In 1948, Gelfond obtained the same value for κ as a corollary of a more general theorem. Finally in 1955, Roth proved the best possible result of $\kappa = 2 + \epsilon$. In Sect. 5.3, we give the proof of Siegel's result as given in Mordell [1] so that the reader can compare and contrast it with the proof of Thue's theorem.

5.1 Approximation Theorem of Dirichlet

Theorem 5.1.1 *Let α and Q be real numbers with $Q > 1$. Then there exist integers p, q such that $1 \leq q < Q$ and $|\alpha q - p| \leq 1/Q$.*

Proof First let us assume that Q is an integer. Consider the following $Q + 1$ numbers

$$0, 1, \{\alpha\}, \{2\alpha\}, \ldots, \{(Q-1)\alpha\}$$

where $\{x\}$ means the fractional part of the real number x. They lie in the unit interval $[0, 1]$. Divide the unit interval into Q subintervals

$$\frac{u}{Q} \leq x < \frac{u+1}{Q}, u \in \{0, 1, \ldots, Q-1\}$$

with $<$ replaced by $\leq$ if $u = Q - 1$. Since there are only Q such intervals, at least one such subinterval contains at least two of the $Q + 1$ numbers listed above. Hence there are integers r_1, r_2, s_1, s_2 with $0 \leq r_i < Q, i = 1, 2$ and $r_1 \neq r_2$ such that

$$|(r_1\alpha - s_1) - (r_2\alpha - s_2)| \leq 1/Q.$$

Taking $q = |r_1 - r_2|$, $p = s_1 - s_2$ or $p = s_2 - s_1$ according as $r_1 > r_2$ or $r_1 < r_2$, respectively, we get

$$1 \leq q < Q \text{ and } |q\alpha - p| \leq 1/Q.$$

Suppose Q is not an integer, then let $Q' = [Q] + 1 > Q$ and apply the above result with Q replaced by Q'. Then $1 \le q < Q'$ implies that $1 \le q \le [Q]$, and hence $1 \le q < Q$. Further $|q\alpha - p| \le 1/Q'$ implies $|q\alpha - p| < 1/Q$ since $Q' > Q$. $\square$

It follows from the above theorem that

$$\left| \alpha - \frac{p}{q} \right| \le \frac{1}{Qq} < \frac{1}{q^2}.$$

In fact there exist infinitely many coprime integers p, q with this property if α is irrational as shown in the corollary below.

Corollary 5.1.2 *Suppose α is irrational. Then there exist infinitely many pairs p, q of relatively prime integers with*

$$\left| \alpha - \frac{p}{q} \right| < \frac{1}{q^2}.$$

Proof Suppose $p = dp', q = dq'$, with $\gcd(p', q') = 1$. Then

$$\left| \alpha - \frac{p'}{q'} \right| = \left| \alpha - \frac{p}{q} \right| < \frac{1}{q^2} \le \frac{1}{q'^2}.$$

Hence in Theorem 5.1.1 we may take p and q as coprime. Since α is irrational, $q\alpha - p$ is never zero. Hence for any given p, q setting $Q_0 = |q\alpha - p|^{-1}$, the inequality $|q\alpha - p| < 1/Q$ can be satisfied only when $Q \le Q_0$. Hence as $Q \to \infty$ there will be infinitely many distinct pairs p, q with $\gcd(p, q) = 1$ satisfying the inequality in Theorem 5.1.1. $\square$

Remark
The above corollary is not true if α is rational. For, then let $\alpha = u/v$ and if $\alpha \ne p/q$, then

$$\left| \alpha - \frac{p}{q} \right| = \left| \frac{u}{v} - \frac{p}{q} \right| = \left| \frac{qu - pv}{vq} \right| \ge \frac{1}{vq}.$$

Hence the inequality in Corollary 5.1.2 is satisfied only if $q < v$. Thus there are only finitely many q values and since

$$|p| < (1 + |\alpha|)|q|$$

there are only finitely many p values satisfying the inequality in the corollary.

In view of the corollary above, we define an irrational number α as *badly approximable* if there is a constant $c = c(\alpha) > 0$ such that

$$\left| \alpha - \frac{p}{q} \right| > \frac{c}{q^2}$$

for every rational p/q.

It can be shown that c must satisfy $0 < c < 1/\sqrt{5}$. This is due to Hurwitz. Further if α is a quadratic irrational, then by a well-known result of Legendre we know that

$$\left| \alpha - \frac{p_n}{q_n} \right| < \frac{1}{q_n^2}$$

for all $n \geq 0$ where p_n/q_n is the nth convergent in the continued fraction expansion of α. Using the best approximation properties of these convergents, one can show that every quadratic irrational is badly approximable. We state here without proof a beautiful result of Khintchine from 1926 which forms basis for metrical transcendence theory.

Theorem 5.1.3 *Suppose $\psi(q)$ is a positive, non-increasing function defined for $q = 1, 2, \ldots$. Consider the inequality*

$$\left| \alpha - \frac{p}{q} \right| < \frac{\psi(q)}{q} \tag{5.2}$$

and the sum

$$\sum_{q=1}^{\infty} \psi(q).$$

If the sum is convergent, then (5.2) has only finitely many solutions in rationals p/q with $q > 0$ for almost all α (in the sense of Lebesgue measure). If the sum is divergent, then (5.2) has infinitely many solutions for almost all α.

As a result of the above theorem, we find that for every $\delta > 0$, the inequality

$$\left| \alpha - \frac{p}{q} \right| < \frac{1}{q^{2+\delta}} \tag{5.3}$$

has only finitely many solutions for almost all α, but

$$\left| \alpha - \frac{p}{q} \right| < \frac{1}{q^2 \log q} \tag{5.4}$$

has infinitely many solutions for almost all α. Since quadratic irrationals are badly approximable, they behave like almost every algebraic number with respect to (5.3), but not with respect to (5.4).

Some Reductions

In the problem of approximating a number $\alpha \in \mathbb{C}$ by rationals, note that if there are only finitely many rationals satisfying the inequality, say

$$\left| \alpha - \frac{p}{q} \right| < \frac{1}{q^{\kappa}}, \kappa \geq 1, \tag{5.5}$$

then there exists a number $C(\alpha, \kappa)$ such that

$$\left|\alpha - \frac{p}{q}\right| > \frac{C(\alpha, \kappa)}{q^{\kappa}} \tag{5.6}$$

for all p/q. Thus in order to prove a result of the type (5.6), we need to prove that (5.5) has only finitely many solutions. Note that when α is a rational number say, s/t, then obviously we have

$$\left|\alpha - \frac{p}{q}\right| > \frac{1}{qt}$$

for any $p/q \neq \alpha$. Hence (5.6) holds. So we will consider only $\alpha \notin \mathbb{Q}$.

While dealing with the approximation problem, we observe that the following simplifications can be made.

(a) We may assume that $\alpha \in \mathbb{R}$. For, otherwise, suppose $\alpha = a + ib$ with $b \neq 0$. Then

$$\left|\alpha - \frac{p}{q}\right| \geq |b| \geq \frac{|b|}{q^{\kappa}}$$

for any p/q.

(b) By similar argument as in (a), we may also assume that

$$\left|\alpha - \frac{p}{q}\right| \leq 1$$

for any p/q. This implies that if q is bounded, then $|p|$ is also bounded since $|p| \leq |q|(1 + |\alpha|)$.

(c) Suppose α is an algebraic number. Then, we may assume that α is an algebraic integer. For otherwise, let d be the denominator of α. Then

$$\left|\alpha - \frac{p}{q}\right| = \frac{1}{d}\left|d\alpha - \frac{dp}{q}\right|.$$

Hence if (5.6) holds for algebraic integers, then the above equality implies that

$$\left|\alpha - \frac{p}{q}\right| \geq \frac{C(\alpha, \kappa)}{dq^{\kappa}} = \frac{C'(\alpha, \kappa)}{q^{\kappa}}$$

where $C'(\alpha, \kappa) = C(\alpha, \kappa)/d$.

(d) Suppose (5.6) is true for all reduced rationals. Let $p = fp'$ and $q = fq'$ with $f \geq 1$ and $\gcd(p', q') = 1$. Then

$$\left|\alpha - \frac{p}{q}\right| = \left|\alpha - \frac{p'}{q'}\right| \geq \frac{C(\alpha, \kappa)}{q'^{\kappa}} = \frac{C(\alpha, \kappa)f^{\eta}}{q^{\kappa}} \geq \frac{C(\alpha, \kappa)}{q^{\kappa}}.$$

Thus it is enough to prove (5.6) for all reduced fractions p/q.

(e) As observed earlier, in order to prove (5.6), we may assume that (5.5) has infinitely many solutions in reduced rationals p/q.

We assume from now onwards that (a)–(e) hold.

5.2   Theorems of Liouville and Thue

Theorem 5.2.1 *Let α be an algebraic number of degree $n \geq 2$. There exists a number $c_{5.1} = c_{5.1}(\alpha) > 0$ such that*

$$\left| \alpha - \frac{p}{q} \right| > \frac{c_{5.1}}{q^n} \tag{5.7}$$

for any integers p, q with $q > 0$.

Proof Let $f(X) = a_0 X^n + \cdots + a_n, a_0 \neq 0$ be the minimal polynomial of α. Then by mean value theorem,

$$\left| f\left(\frac{p}{q} \right) \right| = \left| f(\alpha) - f\left(\frac{p}{q} \right) \right| = \left| f'(\psi)\left(\alpha - \frac{p}{q} \right) \right| \tag{5.8}$$

where ψ is some number between α and p/q. Note that

$$|f'(\psi)| \leq nH(f)(1 + |\psi| + \cdots + |\psi^n|) \leq c_{5.2}$$

since $|\psi| \leq 1 + |\alpha|$ as $|\alpha - p/q| < 1$. Here $c_{5.2}$ denotes a number depending only on α. Thus

$$\left| f\left(\frac{p}{q} \right) \right| < c_{5.2} \left| \alpha - \frac{p}{q} \right|.$$

Further, $f(p/q)$ is non-zero since α is of degree ≥ 2 and

$$\left| f\left(\frac{p}{q} \right) \right| = \frac{|a_0 p^n + a_1 p^{n-1} q + \cdots + a_n q^n|}{q^n} \geq \frac{1}{q^n}. \tag{5.9}$$

Combining (5.8) and (5.9) we get

$$\left| \alpha - \frac{p}{q} \right| > \frac{\min(1, c_{5.2}^{-1})}{q^n}. \qquad \square$$

Theorem 5.2.1 enables one to construct transcendental numbers. For example,

$$L(\mu) = \sum_{j=0}^{\infty} \frac{1}{\mu^{j!}}$$

is transcendental for any integer $\mu > 1$. This is seen as follows. Suppose the number $L(\mu)$ is algebraic of degree $n \geq 1$. For any integer $m \geq 1$, writing

$$\frac{p_m}{q_m} = \sum_{j=0}^{m} \frac{1}{\mu^{j!}}$$

with $\gcd(p_m, q_m) = 1$, we see that $q_m = \mu^{m!}$. Then

$$\left| L(\mu) - \frac{p_m}{q_m} \right| = \sum_{j=m+1}^{\infty} \frac{1}{\mu^{j!}} < \frac{2}{q_m^{n+1}} \text{ for all } m \geq n.$$

Hence $L(\mu)$ cannot be algebraic of degree n, thus showing that $L(\mu)$ is transcendental.

In Theorem 5.2.1, if the exponent n in (5.7) can be reduced, then one can show that certain equations have only finitely many solutions. In 1909, Thue was able to reduce the exponent from n to $n/2 + 1 + \epsilon, \epsilon > 0$. This enabled him to show the following result.

Let $F(X, Y) = a_0 X^n + a_1 X^{n-1} Y + \cdots + a_n Y^n \in \mathbb{Z}[X, Y]$ be an irreducible binary form of degree $n \geq 3, a_0 \neq 0$. Consider equations of the form

$$F(x, y) = k \qquad (5.10)$$

in integers x and y for any fixed integer k. Then the theorem of Thue is as follows.

Theorem 5.2.2 *Equation* (5.10) *has only finitely many solutions.*

Such equations are called *Thue equations*. This theorem is a consequence of the following result.

Theorem 5.2.3 *Let α be an algebraic number of degree $n \geq 3$ and $\epsilon > 0$. Let $\kappa = \frac{n}{2} + 1 + \epsilon$. Then there exists a number $c_{5.3} = c_{5.3}(\alpha, \epsilon) > 0$ such that*

$$\left| \alpha - \frac{p}{q} \right| > \frac{c_{5.3}}{q^{\kappa}}$$

for any $p, q \in \mathbb{Z}, q > 0$.

Derivation of Theorem 5.2.2 from Theorem 5.2.3
Denote by $c_{5.4}, c_{5.5}$ positive numbers depending only on F and k. Write

$$F(X, Y) = a_0(X - \alpha_1 Y) \cdots (X - \alpha_n Y)$$

where $\alpha_1, \ldots, \alpha_n$ are algebraic numbers. They are distinct since F is irreducible. Suppose (5.10) has a solution in integers x and y. We assume that $|x| \leq |y|$, the other case is similar. Let $\alpha \in \{\alpha_1, \ldots, \alpha_n\}$ be such that

$$\left| \alpha - \frac{x}{y} \right| = \min_{1 \le i \le n} \left| \alpha_i - \frac{x}{y} \right|.$$

Let

$$c_{5.4} = \min_{i \ne j} \left| \alpha_i - \alpha_j \right|.$$

First assume that

$$\left| \alpha - \frac{x}{y} \right| < \frac{c_{5.4}}{2}.$$

Then for $\alpha_i \ne \alpha$, we have

$$\left| \alpha_i - \frac{x}{y} \right| \ge |\alpha - \alpha_i| - \left| \alpha - \frac{x}{y} \right| > \frac{c_{5.4}}{2}.$$

From (5.10) we have

$$\prod_{i=1}^{n} \left| \alpha_i - \frac{x}{y} \right| = \left| \frac{k}{a_0 y^n} \right|. \tag{5.11}$$

Hence

$$\left| \alpha - \frac{x}{y} \right| < \frac{c_{5.5}}{|y|^n}$$

where

$$c_{5.5} = |k| 2^{n-1} / (|a_0| c_{5.4}^{n-1}).$$

By Theorem 5.2.3, we get

$$\frac{c_{5.3}}{|y|^\kappa} < \frac{c_{5.5}}{|y|^n}$$

which gives

$$|y|^{n-\kappa} < c_{5.5} c_{5.3}^{-1}.$$

By taking $\epsilon < n/2 - 1$, the above inequality implies that there are only finitely many values for y and hence for x since $|x| \le |y|$.

Next if $\left| \alpha - \frac{x}{y} \right| \ge c_{5.4}/2$, then from (5.11), we get

$$|y|^n \le |k| 2^n / (|a_0| c_{5.4}^n)$$

giving the desired result as before.   □

Proof of Theorem 5.2.3

The proof is based on a paper by Davenport [2]. Let the assumptions (a)–(e) in Sect. 6.1 hold.

Step 1. Construction of an auxiliary polynomial which vanishes at (α, α) to a high order

Let L, k be parameters to be chosen later. Put

$$R(X, Y) = \sum_{\lambda_1=0}^{L} \sum_{\lambda_2=0}^{1} p(\lambda_1, \lambda_2) X^{\lambda_1} Y^{\lambda_2} \in \mathbb{Z}[X, Y] \tag{5.12}$$

and

$$R_m(X, Y) = \frac{1}{m!} \frac{\partial^m}{\partial X^m} R(X, Y) \tag{5.13}$$

where $p(\lambda_1, \lambda_2)$ are integral variables to be determined subject to the condition

$$R_m(\alpha, \alpha) = 0 \text{ for } 0 \leq m < k. \tag{5.14}$$

We have

$$R_m(\alpha, \alpha) = \sum_{\lambda_1=0}^{L} \sum_{\lambda_2=0}^{1} p(\lambda_1, \lambda_2) \binom{\lambda_1}{m} \alpha^{\lambda_1 + \lambda_2 - m}$$

which by Lemma 1.3.1 can be written as

$$R_m(\alpha, \alpha) = \sum_{j=0}^{n-1} \alpha^j \sum_{\lambda_1=0}^{L} \sum_{\lambda_2=0}^{1} p(\lambda_1, \lambda_2) \binom{\lambda_1}{m} b_{j,\lambda_1+\lambda_2-m}$$

and

$$\max_{(\lambda_1,\lambda_2)} \left| \binom{\lambda_1}{m} b_{j,\lambda_1+\lambda_2-m} \right| < 2^L (2h(\alpha))^{L+1} \leq (4h(\alpha))^{2L} \text{ for } m \geq 0.$$

Since α is of degree n, the system of equations in (5.14) is equivalent to

$$\sum_{\lambda_1=0}^{L} \sum_{\lambda_2=0}^{1} p(\lambda_1, \lambda_2) \binom{\lambda_1}{m} b_{j,\lambda_1+\lambda_2-m} = 0, 0 \leq m < k, 0 \leq j \leq n-1.$$

There are nk equations in $2(L+1)$ unknowns $p(\lambda_1, \lambda_2)$. We shall choose

$$L = \left[\frac{1}{2}(1 + \delta)nk \right] \tag{5.15}$$

so that $2(L+1) > (1+\delta)nk, \delta > 0$. Then by Lemma 3.1.1, there exist $p(\lambda_1, \lambda_2) \in \mathbb{Z}$, not all zero, such that

$$|p(\lambda,\lambda_2)| \le (4(L+1)(4h(\alpha))^{2L})^{1/\delta} \le (4h(\alpha))^{4L/\delta}. \qquad (5.16)$$

Step 2. An upper bound for $\left|R_m\left(\frac{p_1}{q_1}, \frac{p_2}{q_2}\right)\right|$ where $\frac{p_1}{q_1}$ and $\frac{p_2}{q_2}$ are two rationals very close to α

Denote by $c_{5.6}, c_{5.7}, \ldots$ positive numbers depending only on α and δ. Let p_1/q_1 and p_2/q_2 be two reduced rational numbers with

$$\left|\alpha - \frac{p_i}{q_i}\right| < 1, i = 1, 2.$$

Then $|p_i/q_i| < 1 + |\alpha|, i = 1, 2$. By (5.12), we can write

$$R(X, Y) = P(X) - YQ(X)$$

where

$$P(X) = \sum_{\lambda_1=0}^{L} p(\lambda_1, 0)X^{\lambda_1}; \quad Q(X) = -\sum_{\lambda_1=0}^{L} p(\lambda_1, 1)X^{\lambda_1}.$$

Then by the definition of $R_m(X, Y)$, it follows that

$$R_m(X, Y) = P_m(X) - YQ_m(X)$$

where

$$P_m(X) = \frac{1}{m!}P^{(m)}(X) = \sum_{\lambda_1=0}^{L} p(\lambda_1, 0)\binom{\lambda_1}{m}X^{\lambda_1-m}$$

and

$$Q_m(X) = \frac{1}{m!}Q^{(m)}(X) = \sum_{\lambda_1=0}^{L} p(\lambda_1, 1)\binom{\lambda_1}{m}X^{\lambda_1-m}.$$

Applying (5.16) we get

$$|P_m(X)| \le (L+1)(4h(\alpha))^{4L/\delta}2^L(1+|X|)^L \le (c_{5.6}(1+|X|))^L$$

where $c_{5.6}$ may be taken as $4(4h(\alpha))^{4/\delta}$. Similarly

$$|Q_m(X)| \le (c_{5.6}(1+|X|))^L. \qquad (5.17)$$

Thus

$$|R_m(X, Y)| \le |P_m(X)| + |Y||Q_m(X)| \le (c_{5.6}(1+|X|))^L(1+|Y|). \qquad (5.18)$$

Taking $X = p_1/q_1, Y = p_2/q_2$ and observing that

$$R_m\left(\frac{p_1}{q_1}, \frac{p_2}{q_2}\right) = P_m\left(\frac{p_1}{q_1}\right) - \alpha Q_m\left(\frac{p_1}{q_1}\right) - \left(\frac{p_2}{q_2} - \alpha\right) Q_m\left(\frac{p_1}{q_1}\right),$$

we obtain

$$\left|R_m\left(\frac{p_1}{q_1}, \frac{p_2}{q_2}\right)\right| \le \left|R_m\left(\frac{p_1}{q_1}, \alpha\right)\right| + \left|\alpha - \frac{p_2}{q_2}\right| \left|Q_m\left(\frac{p_1}{q_1}\right)\right|$$

which by (5.17) implies that

$$\left|R_m\left(\frac{p_1}{q_1}, \frac{p_2}{q_2}\right)\right| \le \left|R_m\left(\frac{p_1}{q_1}, \alpha\right)\right| + \left|\alpha - \frac{p_2}{q_2}\right| c_{5.7}^L. \tag{5.19}$$

Estimate for $\left|R_m\left(\frac{p_1}{q_1}, \alpha\right)\right|$

Fix an integer m with $0 \le m < k$. Let

$$S(X) = R_m(X, \alpha) \text{ and } S_\nu(\alpha) = \frac{1}{\nu!} S^{(\nu)}(\alpha).$$

Then by (5.13),

$$S_\nu(\alpha) = \binom{m + \nu}{\nu} R_{m+\nu}(\alpha, \alpha).$$

Using (5.14), we get

$$S_\nu(\alpha) = 0 \text{ for } \nu < k - m.$$

Hence

$$S\left(\frac{p_1}{q_1}\right) = S\left(\frac{p_1}{q_1} - \alpha + \alpha\right) = \sum_{\nu=k-m}^{\infty} S_\nu(\alpha) \left(\frac{p_1}{q_1} - \alpha\right)^\nu$$

by Taylor expansion, thus giving

$$R_m\left(\frac{p_1}{q_1}, \alpha\right) = \sum_{\nu=k-m}^{L-m} \binom{m + \nu}{\nu} R_{m+\nu}(\alpha, \alpha) \left(\frac{p_1}{q_1} - \alpha\right)^\nu.$$

By $\left|\alpha - \frac{p_1}{q_1}\right| < 1$ and (5.18), we therefore get

$$\left|R_m\left(\frac{p_1}{q_1}, \alpha\right)\right| \le c_{5.8}^L \left|\frac{p_1}{q_1} - \alpha\right|^{k-m}.$$

Combining this with (5.19) we obtain

$$\left| R_m \left(\frac{p_1}{q_1}, \frac{p_2}{q_2} \right) \right| \leq \left(\left| \alpha - \frac{p_1}{q_1} \right|^{k-m} + \left| \alpha - \frac{p_2}{q_2} \right| \right) c_{5.9}^L. \qquad (5.20)$$

Step 3. There exists some integer m with $R_m \left(\frac{p_1}{q_1}, \frac{p_2}{q_2} \right) \neq 0$

Let t be a positive integer such that

$$R_m \left(\frac{p_1}{q_1}, \frac{p_2}{q_2} \right) = 0 \text{ for } 0 \leq m \leq t.$$

We shall get an upper bound T for t. Then there exists some m with $0 \leq m \leq T+1$ such that $R_m \left(\frac{p_1}{q_1}, \frac{p_2}{q_2} \right) \neq 0$.

Let us take two distinct integers m and m' with $0 \leq m, m' \leq t$. From the definition of R_m it is clear that

$$P^{(m)} \left(\frac{p_1}{q_1} \right) - \frac{p_2}{q_2} Q^{(m)} \left(\frac{p_1}{q_1} \right) = 0$$

and

$$P^{(m')} \left(\frac{p_1}{q_1} \right) - \frac{p_2}{q_2} Q^{(m')} \left(\frac{p_1}{q_1} \right) = 0.$$

Eliminating p_2/q_2 in these equations, we get

$$P^{(m)} \left(\frac{p_1}{q_1} \right) Q^{(m')} \left(\frac{p_1}{q_1} \right) - P^{(m')} \left(\frac{p_1}{q_1} \right) Q^{(m)} \left(\frac{p_1}{q_1} \right) = 0.$$

Put

$$W(X) = P(X)Q^{(1)}(X) - P^{(1)}(X)Q(X).$$

Then $W(X) \in \mathbb{Z}[X]$ and by the previous identity, we get

$$W^{(\mu)} \left(\frac{p_1}{q_1} \right) = 0 \text{ for } 0 \leq \mu < t.$$

Therefore,

$$W(X) = (q_1 X - p_1)^t U(X) \text{ with } U(X) \in \mathbb{Z}[X] \qquad (5.21)$$

by gcd $(p_1, q_1) = 1$ and Gauss Lemma 1.2.1. By Step 1, it follows that $H(W)$, the height of W satisfies

$$H(W) \leq c_{5.10}^L \leq c_{5.11}^k. \qquad (5.22)$$

Suppose $W(X) \not\equiv 0$. Then by (5.21), $H(W) \geq q_1^t$. Comparing this with (5.22), we get

$$t \leq \left\lceil \frac{k \log c_{5.11}}{\log q_1} \right\rceil.$$

Thus we may take

$$T = \left\lceil \frac{k \log c_{5.11}}{\log q_1} \right\rceil. \tag{5.23}$$

Now we show that $W(X) \not\equiv 0$. By Step 1, we observe that either $P(X)$ or $Q(X)$ is not identically zero. Suppose $Q(X) \equiv 0$. Then $P(X) \not\equiv 0$. Again by Step 1,

$$P^{(m)}(\alpha) = 0 \text{ for } 0 \le m < k$$

which implies

$$P^{(m)}(\alpha^{(j)}) = 0 \text{ for } 0 \le m < k, 1 \le j \le n$$

where $\alpha^{(1)} = \alpha, \ldots, \alpha^{(n)}$ are the conjugates of α. Since $P(X) \not\equiv 0$ we have

$$nk \le \deg P \le L < nk,$$

by (5.15) and taking $0 < \delta < 1$. This is a contradiction. Thus $Q(X) \not\equiv 0$. Next suppose that $P(X)/Q(X)$ is a constant function, say λ. Then $\lambda \in \mathbb{Q}$. Hence $\lambda \ne \alpha$. Also

$$R(X, \alpha) = (\lambda - \alpha)Q(X).$$

So by Step 1, we get that $Q^{(m)}(\alpha^{(j)}) = 0$ for $0 \le m < k, 1 \le j \le n$. By a comparison of the degree, as earlier, we get a contradiction. Thus $P(X)/Q(X)$ is not a constant function. Hence

$$\left(\frac{P(X)}{Q(X)} \right)' = -\frac{W(X)}{Q^2(X)} \not\equiv 0$$

implying $W(X) \not\equiv 0$. In conclusion, there exists an integer m say, m_0 with $0 \le m_0 \le T + 1$ for which $R_{m_0}\left(\frac{p_1}{q_1}, \frac{p_2}{q_2}\right) \ne 0$ with T as in (5.23).

Step 4. A lower bound for $\left| R_{m_0}\left(\frac{p_1}{q_1}, \frac{p_2}{q_2}\right) \right|$

Among the infinitely many rationals satisfying (5.5) we will choose suitably two rationals p_1/q_1 and p_2/q_2 satisfying $q_2 > q_1 > 1$. Let L be given by (5.15) and take $0 < \delta < 1/12$ so that Steps 1–3 are valid. Take

$$k = \left\lfloor \frac{\log q_2}{\log q_1} \right\rfloor$$

and

$$q_1 > e^{\log c_{5.11}/\delta}.$$

Then

$$q_1^k \le q_2 < q_1^{k+1}. \tag{5.24}$$

and

$$m_0 \leq \left\lceil \frac{k \log c_{5.11}}{\log q_1} \right\rceil + 1 \leq k\delta + 2. \tag{5.25}$$

From the definition of R_m and (5.24), it is easy to see that

$$\left| R_{m_0}\left(\frac{p_1}{q_1}, \frac{p_2}{q_2} \right) \right| \geq q_1^{-(1+\delta)nk/2 - k - 1}. \tag{5.26}$$

Step 5. An upper bound for $\left| R_{m_0}\left(\frac{p_1}{q_1}, \frac{p_2}{q_2} \right) \right|$

By (5.20), assumption (5.5) and (5.25), we get

$$\left| R_{m_0}\left(\frac{p_1}{q_1}, \frac{p_2}{q_2} \right) \right| \leq \max(q_1^{\kappa(-k(1-\delta)+2)}, q_2^{-\kappa}) c_{5.12}^k.$$

Let us now take $q_1 > \max(e^{\log c_{5.11}/\delta}, c_{5.12}^{1/\delta})$. Then Step 4 is valid and also

$$\left| R_{m_0}\left(\frac{p_1}{q_1}, \frac{p_2}{q_2} \right) \right| \leq q_1^{\kappa(-k(1-\delta)+2)+\delta k}. \tag{5.27}$$

Step 6. Gap principle

For any two reduced rationals P_1/Q_1 and P_2/Q_2 with $Q_2 > Q_1 > 1$ satisfying (5.5) we see that

$$\frac{1}{Q_1 Q_2} \leq \left| \frac{P_1}{Q_1} - \frac{P_2}{Q_2} \right| \leq \left| \alpha - \frac{P_1}{Q_1} \right| + \left| \alpha - \frac{P_2}{Q_2} \right| < \frac{2}{Q_1^\kappa}$$

which gives

$$Q_2 > Q_1^{\kappa-1}/2.$$

In other words, denominators of such rationals are far apart. This phenomenon is also described as *good approximations repel one another.*

Step 7. Final contradiction

Take a rational p_2/q_2 with $q_2 > 0$ satisfying (5.5) and so large that

$$k = \left\lceil \frac{\log q_2}{\log q_1} \right\rceil > \delta^{-1}.$$

This is possible by Step 6. Comparing (5.26) and (5.27), we get

$$\kappa \leq \frac{(1+\delta)k(n/2+1)+1}{k(1-\delta)-2}$$

$$\leq \frac{n}{2} + 1 + \frac{(2\delta k + 2)(n/2+1)+1}{k(1-\delta)-2}.$$

Since $\delta < 1/12$, we estimate the last term in the above inequality as

$$< \frac{12}{11k - 24}(2\delta k + 3)(n/2 + 1) < 7\delta(n/2 + 1).$$

Taking $\delta < \epsilon/(4n + 7)$, we get $\kappa < n/2 + 1 + \epsilon$, a contradiction. □

Remark

The proof of Thue's result depends on the existence of a solution p_1/q_1 for (5.5) with large q_1. This makes the proof *ineffective*. This continues in the proof of Siegel's theorem below.

5.3 Theorem of Siegel

Theorem 5.3.1 *Let α be an algebraic number of degree $n \geq 3$ and $\epsilon > 0$. There exists a number $c_{5.13} = c_{5.13}(\alpha, \epsilon) > 0$ such that*

$$\left| \alpha - \frac{p}{q} \right| > \frac{c_{5.13}}{q^\kappa}$$

for any $p, q \in \mathbb{Z}, q > 0$ provided κ is as in (5.1).

Proof We denote by $c_{5.14}, \ldots$ positive numbers depending only on α and some parameter $\delta > 0$ which will be chosen later.

We need to generalise several ideas in the proof of Theorem 5.2.2. As before, we do them in several steps. Let all the assumptions (a)–(e) of Sect. 6.1 hold.

Step 1′. Construction of an auxiliary polynomial which vanishes at (α, α) to a high order

Note that in Theorem 5.2.2 the polynomial $R(X, Y)$ is linear in Y. Now we shall construct a nonlinear polynomial in Y. Let a, b, r be non-negative integers. Put

$$R(X, Y) = \sum_{\lambda_1=0}^{a} \sum_{\lambda_2=0}^{b} p(\lambda_1, \lambda_2) X^{\lambda_1} Y^{\lambda_2} \in \mathbb{Z}[X, Y]$$

and

$$R_m(X, Y) = \frac{1}{m!} \frac{\partial^m}{\partial X^m} R(X, Y)$$

where $p(\lambda_1, \lambda_2)$ are integral variables to be determined subject to the condition

$$R(X, \alpha) = (X - \alpha)^r S(X) \tag{5.28}$$

for some polynomial S and

$$R_m(\alpha, \alpha) = 0 \text{ for } 0 \le m < r. \tag{5.29}$$

The vanishing of any polynomial $P(X) \in \mathbb{Q}[X]$ at $X = \alpha$ requires n conditions given by expressing the powers of α in $P(\alpha)$ in terms of $1, \alpha, \dots, \alpha^{n-1}$, and then equating to zero, the coefficients of these powers. We also have

$$P(\alpha_1) = \cdots = P(\alpha_n) = 0$$

where $\alpha = \alpha_1, \dots, \alpha_n$ are the conjugates of α. This is equivalent to saying,

$$\alpha_1^t P(\alpha_1) + \cdots + \alpha_n^t P(\alpha_n) = 0, 0 \le t < n.$$

Taking $P(X) = R_m(X, \alpha)$ each condition in (5.29) gives rise to n conditions on the coefficients $p(\lambda_1, \lambda_2)$ in

$$\alpha_1^t R_m(\alpha_1, \alpha_1) + \cdots + \alpha_n^t R_m(\alpha_n, \alpha_n) = 0. \tag{5.30}$$

Thus there are rn conditions for $(a + 1)(b + 1)$ unknowns p_{λ_1, λ_2}. Note also that the coefficients of each p_{λ_1, λ_2} are in $\mathbb{Z}$.

Now we choose the parameters a, b as follows.

$$0 < b < n \text{ and } a = \left[\left(\frac{n + \delta}{b + 1}\right) r\right].$$

Then

$$(a + 1)(b + 1) - nr > \delta r. \tag{5.31}$$

Note that

$$R_m(\alpha_i, \alpha_i) = \sum_{\lambda_1 = m}^{a} \sum_{\lambda_2 = 0}^{b} \binom{\lambda_1}{m} p(\lambda_1, \lambda_2) \alpha_i^{\lambda_1 + \lambda_2 - m}, 1 \le i \le n.$$

Hence the coefficient of $p(\lambda_1, \lambda_2)$ in $R_m(\alpha_i, \alpha_i)$ has absolute value bounded by

$$2^a \max(1, |\alpha_i|^{a+b}) < c_{5.14}^r.$$

Thus from (5.30) we get that the coefficients of $p(\lambda_1, \lambda_2)$ are bounded in absolute value by

$$c_{5.14}^r(|\alpha_1|^t + \cdots + |\alpha_n|^t) \le c_{5.15}^r.$$

Applying Lemma 3.1.1, there exist rational integers $p(\lambda_1, \lambda_2)$, not all zero, such that

$$\max |p(\lambda_1, \lambda_2)| \le \left((a + 1)(b + 1)c_{5.15}^r\right)^{nr/((a+1)(b+1)-nr)}.$$

By (5.31) and since

$$(a+1)(b+1)c_{5.15}^r < \left(\frac{n+1}{b+1}r+1\right)(b+1)c_{5.15}^r < (n+1)(r+1)c_{5.15}^r,$$

we find that there exist $p(\lambda_1, \lambda_2)$, not all zero and satisfying (5.29) with

$$|p(\lambda_1, \lambda_2)| < ((n+1)(r+1)c_{5.15}^r)^{n/\delta} < c_{5.16}^r.$$

From this we find that the coefficients of powers of α in the expression for $R_m(\alpha, \alpha)$ are bounded by $c_{5.17}^r$.

Step 2′. An upper bound for $\left|R_m\left(\frac{p_1}{q_1}, \frac{p_2}{q_2}\right)\right|$ where $\frac{p_1}{q_1}$ and $\frac{p_2}{q_2}$ are two rationals very close to α

Let p_1/q_1 and p_2/q_2 be two reduced rational numbers with

$$\left|\alpha - \frac{p_i}{q_i}\right| < 1, i = 1, 2$$

and let $t_i = \dfrac{p_i}{q_i} - \alpha, i = 1, 2$. Write $R_m(X, Y)$ as

$$R_m(X, Y) = \sum_{\lambda_1=0}^{a-m} \sum_{\lambda_2=0}^{b} q_{\lambda_1, \lambda_2} X^{\lambda_1} Y^{\lambda_2}.$$

Then by Step 1′,

$$|q_{\lambda_1, \lambda_2}| \leq c_{5.17}^r.$$

Further

$$R_m\left(\frac{p_1}{q_1}, \frac{p_2}{q_2}\right) = \sum_{\lambda_1=0}^{a-m} \sum_{\lambda_2=0}^{b} q_{\lambda_1, \lambda_2} (t_1 + \alpha)^{\lambda_1} (t_2 + \alpha)^{\lambda_2} = \sum_{\lambda_1=0}^{a-m} \sum_{\lambda_2=0}^{b} s_{\lambda_1, \lambda_2} t_1^{\lambda_1} t_2^{\lambda_2}$$

where

$$|s_{\lambda_1, \lambda_2}| \leq \sum_{\lambda_1=0}^{a-m} \sum_{\lambda_2=0}^{b} |q_{\lambda_1, \lambda_2}| (1 + |\alpha|)^{\lambda_1} (1 + |\alpha|)^{\lambda_2}$$

$$\leq (a+1)(b+1)c_{5.17}^r (1 + |\alpha|)^{a+b} \leq c_{5.18}^r.$$

Note that

$$s_{\lambda_1, 0} = 0 \text{ for } 0 \leq \lambda_1 \leq r - 1 - m.$$

Hence

$$R_m\left(\frac{p_1}{q_1}, \frac{p_2}{q_2}\right) = t_1^{r-m} \sum_{\lambda_1=r-m}^{a-m} s_{\lambda_1,0} t_1^{\lambda_1-r+m} + t_2 \sum_{\lambda_1=0}^{a-\lambda_1} \sum_{\lambda_2=1}^{b} s_{\lambda_1,\lambda_2} t_1^{\lambda_1} t_2^{\lambda_2-1}.$$

Thus using $|t_1| < 1$ and $|t_2| < 1$, we get

$$\left| R_m\left(\frac{p_1}{q_1}, \frac{p_2}{q_2}\right) \right| \leq |t_1|^{r-m} \sum_{\lambda_1=r-m}^{a-m} c_{5.18}^r + |t_2| \sum_{\lambda_1=0}^{a-\lambda_1} \sum_{\lambda_2=1}^{b} c_{5.18}^r$$

$$\leq |t_1|^{r-m}(a-r+1)c_{5.18}^r + |t_2|(a+1)(b+1)c_{5.18}^r$$

$$\leq c_{5.19}^r(|t_1|^{r-m} + |t_2|).$$

Therefore we get

$$q_1^a q_2^b \left| R_m\left(\frac{p_1}{q_1}, \frac{p_2}{q_2}\right) \right| \leq c_{5.19}^r \left(\left| \alpha - \frac{p_1}{q_1} \right|^{r-m} + \left| \alpha - \frac{p_2}{q_2} \right| \right) q_1^a q_2^b. \qquad (5.32)$$

Step 3′. There exists some integer m, $0 \leq m < r$ with $R_m\left(\frac{p_1}{q_1}, \frac{p_2}{q_2}\right) \neq 0$
This step is the most difficult part in the proof, and it is more complicated than Step 3 in Theorem 5.2.2.

We proceed as follows. Write

$$R(X, Y) = \sum_{\lambda_2=0}^{b} f_{\lambda_2}(X) Y^{\lambda_2} = \sum_{\lambda_1=0}^{a} g_{\lambda_1}(Y) X^{\lambda_1}$$

where $f_{\lambda_2}(X)$ and $g_{\lambda_1}(Y)$ are polynomials in X and Y, respectively. By construction in Step 1′ we know that not all the polynomials $f_{\lambda_2}(X)$ and not all the polynomials $g_{\lambda_1}(Y)$ are identically zero. Among the polynomials $f_0(X), \ldots, f_b(X)$, suppose $b'+1, 0 < b' \leq b$ functions are linearly independent. Let us denote them by

$$F_0(X), F_1(X), \ldots, F_{b'}(X).$$

Then by Lemma 1.4.1,

$$\Delta = ||F_\lambda^\mu(X)|| \neq 0, 0 \leq \mu \leq b'; 0 \leq \lambda \leq b'.$$

Write

$$R(X, Y) = \sum_{\lambda=0}^{b'} F_\lambda(X) G_\lambda(Y)$$

with $F_\lambda(X) \in \mathbb{Z}[X]$ and $G_\lambda(Y) \in \mathbb{Q}[Y]$ for $0 \le \lambda \le b'$. Note that $G_\lambda(Y)$ are also linearly independent since, otherwise, $R(X, Y)$ could be expressed with less than b' of the $F_\lambda(X)$.

We shall show that

$$R\left(X, \frac{p_2}{q_2}\right) \not\equiv 0$$

if q_2 is sufficiently large. Suppose not. Then

$$\sum_{\lambda_1=0}^{a} g_{\lambda_1}\left(\frac{p_2}{q_2}\right) X^{\lambda_1} \equiv 0.$$

Hence for some λ_1 for which $g_{\lambda_1}(Y) \not\equiv 0$, we have

$$g_{\lambda_1}\left(\frac{p_2}{q_2}\right) = 0.$$

Writing $g_{\lambda_1}(Y) = p_{\lambda_1,0} + p_{\lambda_1,1} Y + \cdots + p_{\lambda_1,b''} Y^{b''}$, with $p_{\lambda_1,b''}$ as the largest non-vanishing coefficient, we have

$$\sum_{\lambda_2=0}^{b''} p_{\lambda_1,\lambda_2}\left(\frac{p_2}{q_2}\right)^\lambda = 0.$$

This implies that $q_2 | p_{\lambda_1,b''}$. By Step $1'$, this means

$$q_2 \le c_{5.16}^r.$$

Thus we conclude that

$$R\left(X, \frac{p_2}{q_2}\right) \not\equiv 0 \text{ if } q_2 > c_{5.16}^r. \tag{5.33}$$

From now on we shall assume that $q_2 > c_{5.16}^r$. We now claim that there exists m with $0 \le m < r$ such that

$$R_m\left(h, \frac{p_2}{q_2}\right) \ne 0 \text{ for any } h \in \mathbb{Q}$$

provided

$$\Delta^{(\gamma)}(h) \ne 0 \text{ for some } \gamma \text{ with } b + \gamma < r. \tag{5.34}$$

Assume first that (5.34) holds. We show the claim. By (5.33), we have

$$R\left(X, \frac{p_2}{q_2}\right) = \sum_{\lambda=0}^{b'} F_\lambda(X) G_\lambda\left(\frac{p_2}{q_2}\right) \not\equiv 0.$$

Hence we may assume without loss of generality that $G_0\left(\frac{p_2}{q_2}\right) \neq 0$. Differentiating the above expression repeatedly, we get

$$m! R_m\left(X, \frac{p_2}{q_2}\right) = \sum_{\lambda=0}^{b'} F_\lambda^{(m)}(X) G_\lambda\left(\frac{p_2}{q_2}\right), 0 \le m \le b'.$$

Since the determinant of the coefficient matrix of the above system, $\Delta(X)$ is not identically zero, we can solve for $G_\lambda\left(\frac{p_2}{q_2}\right)$. Thus we have

$$\Delta(X) G_0\left(\frac{p_2}{q_2}\right) = \sum_{\lambda=0}^{b'} H_\lambda(X) R_\lambda\left(X, \frac{p_2}{q_2}\right),$$

where $H_\lambda(X)$ are polynomials with integer coefficients. Differentiating the above expression γ times and putting $X = h$, we obtain

$$\Delta^{(\gamma)}(h) G_0\left(\frac{p_2}{q_2}\right) = \sum_{\lambda=0}^{b'+\gamma} h_\lambda R_\lambda\left(h, \frac{p_2}{q_2}\right)$$

where h_λ are rational numbers. Since $b' + \gamma \le b + \gamma < r$, the R_λ are defined by (5.28) and not all the $R_\lambda\left(h, \frac{p_2}{q_2}\right)$ are zero for $0 \le \lambda \le b' + \gamma$. This proves the claim.

Next we find a γ satisfying (5.34). We have

$$m! R_m(X, \alpha) = \sum_{\lambda=0}^{b'} F_\lambda^{(m)}(X) G_\lambda(\alpha), 0 \le m \le b'.$$

The polynomial $R_m(X, \alpha)$ in X is divisible by $(X - \alpha)^{r-m}$ and so by $(X - \alpha)^{r-b'}$ since $m \le b' < r$. Arguing as in the previous paragraph, we get that $\Delta(X) G_0(\alpha)$ is a linear combination of $R_m(X, \alpha)$ and hence divisible by $(X - \alpha)^{r-m}$. Now $G_0(X) \not\equiv 0$ and $G_0(\alpha) \neq 0$ since $G_0(X)$ is of degree $b < n$. Hence $\Delta(X)$ is divisible by $(X - \alpha)^{r-b'}$. So $\Delta(X)$ is divisible by the minimal polynomial say, $f(X)$ of α to the power $r - b'$. Thus we can write

$$\Delta(X) = f(X)^{r-b'} D(X)$$

where $D(X) \in \mathbb{Q}[X]$ is of degree d, say. We know that the elements of $\Delta(X)$ are of degree at most a. Hence $\Delta(X)$ is of degree at most $a(b' + 1)$. From the above expression for $\Delta(X)$ we get

$$n(r - b') + d \le a(b' + 1).$$

By the choices of a and b we get

$$d \le \left(\frac{n + \delta}{b + 1}\right) r(b + 1) - nr + nb \le \delta r + nb \le \delta r + n^2 - n.$$

Since $h \in \mathbb{Q}$, $f(h) \ne 0$, and so $\Delta(X)$ has a zero at $X = h$ of order γ with

$$\gamma \le \delta r + n^2 - n.$$

Thus (5.34) is satisfied if $b + \delta r + n^2 - n < \delta r + n^2 < r$.

Taking $h = p_1/q_1$ and summarising the above arguments, we get

$$R_m \left(\frac{p_1}{q_1}, \frac{p_2}{q_2}\right) \ne 0 \text{ for some } m \text{ with } 0 \le m < r$$

provided $\delta r + n^2 < r$ and $q_2 > c_{5.16}^r$.

Step 4'. A lower bound for $R_m \left(\frac{p_1}{q_1}, \frac{p_2}{q_2}\right)$

We follow the Step 4 in Theorem 5.2.2. Among the infinitely many rationals satisfying (5.5) we will choose suitably two rationals p_1/q_1 and p_2/q_2 satisfying $q_1^r \le q_2 < q_1^{r+1}$ so that

$$r = \left\lceil \frac{\log q_2}{\log q_1} \right\rceil.$$

Let a and b be as given in Step 1' and take $0 < \delta < 1/2$ and $q_2 > c_{5.16}^r$ so that Steps $1' - 3'$ are valid. Further we take q_2 so large that $r > 2n^2$. Then

$$m \le \delta r + n^2 < r. \tag{5.35}$$

Thus

$$q_1^a q_2^b \left| R_m \left(\frac{p_1}{q_1}, \frac{p_2}{q_2}\right) \right| \ge 1.$$

Together with (5.32), this gives

$$c_{5.19}^r q_1^a q_2^b \max \left(\left| \alpha - \frac{p_1}{q_1} \right|^{r-m}, \left| \alpha - \frac{p_2}{q_2} \right| \right) > 1. \tag{5.36}$$

Step 5'. Final contradiction

We shall choose q_1 and δ in such a way that inequality (5.36) is contradicted. Let

$$\kappa = \frac{n}{b + 1} + b + \epsilon, 0 < \epsilon < 1.$$

Then

$$\kappa < \frac{n}{2} + n - 1 + 1 < 2n.$$

Consider

$$A := q_1^a q_2^b \left| \alpha - \frac{p_1}{q_1} \right|^{r-m} \le q_1^{a+b(r+1)-\kappa(r-m)}$$

by the choice of q_2 and (5.5). Using the value for a and (5.35) we get

$$A \le q_1^{\left(\frac{n+\delta}{b+1} + b + \frac{b}{r} - \kappa + \frac{m\kappa}{r}\right)r} < q_1^{\left(\frac{\delta}{b+1} + \delta\kappa + \frac{b}{r} + \frac{\kappa n^2}{r} - \epsilon\right)r}.$$

Take δ such that

$$\frac{\delta}{b+1} + \delta\kappa < \epsilon/2$$

and q_2 so large that r satisfies

$$\frac{b}{r} + \frac{\kappa n^2}{r} < \epsilon/4.$$

Then

$$A < q_1^{-\epsilon r/4}. \tag{5.37}$$

Next,

$$B := q_1^a q_2^b \left| \alpha - \frac{p_2}{q_2} \right| < q_1^{\frac{n+\delta}{b+1}r} q_2^{b-\kappa} < q_1^{\left(\frac{n+\delta}{b+1} + b - \kappa\right)r}$$

since $b < \kappa$ and $q_2 \ge q_1^r$. Thus

$$B < q_1^{\left(\frac{\delta}{b+1} - \epsilon\right)r} < q_1^{-\epsilon r/2}$$

by the choice of δ. Hence (5.37) is true with A replaced by B. From (5.36), we get

$$1 < c_{5.19}^r \max(A, B) < (c_{5.19} q_1^{-\epsilon/4})^r.$$

Choose q_1 large so that $(c_{5.19} q_1^{-\epsilon/4})^r < 1$. This choice of q_1 contradicts the above inequality giving the final contradiction.

Exercise

(1) Let P be the set of natural numbers all of whose prime divisors belong to a finite set of primes $\{p_1, \ldots, p_m\}$. Show that the equation

$$x - y = k, k \in \mathbb{Z}, k \ne 0$$

has only finitely many solutions in $x, y \in P$,

(2) Let $F(X, Y) = a_0 X^n + a_1 X^{n-1} Y + \cdots + a_n Y^n \in \mathbb{Z}[X, Y]$ be an irreducible binary form of degree $n \geq 3$. Then show that for any $\nu < n - 2\sqrt{n}$, there are only finitely many integer points (x, y) with

$$0 < |F(x, y)| < (|x| + |y|)^\nu.$$

Suppose $G(X, Y)$ is a polynomial of total degree $\nu < n - 2\sqrt{n}$. Then show that there are only finitely many integer points (x, y) with

$$F(x, y) = G(x, y).$$

(3) Let α be an algebraic number of degree n and let $P(X) \in \mathbb{Z}[X]$ be of degree k. Show that there exists a number $c = c(\alpha) > 0$ such that either $P(\alpha) = 0$ or

$$|P(\alpha)| > \frac{c^k}{H(P)^{n-1}}.$$

Notes

Given any real number α, one would like to know whether it behaves like almost every other number. For example, if α is a quadratic irrational, then it is badly approximable, and hence, it behaves like almost every number with respect to inequality (5.3) but not with respect to inequality (5.4). As another example, consider the Euler constant e. It has the continued fraction expansion as

$$[2; 1, 2, 1, 1, 4, 1, 1, 6, 1, \ldots].$$

In 1978, Davis [3] used this continued fraction expansion to obtain the following rational approximation to e. Let $\epsilon > 0$. Then for any $p, q \in \mathbb{N}$ with $q > q_0(\epsilon)$, one has

$$\left| e - \frac{p}{q} \right| > \left(\frac{1}{2} - \epsilon \right) \frac{\log \log q}{q^2 \log q}.$$

From this it is clear that neither of the inequalities (6.4) and (6.5) has infinitely many solutions if $\alpha = e$. Thus with respect to (5.3), the constant e behaves like almost every number but not with respect to (5.4). In 1953, Mahler [4] showed that

$$\left| \pi - \frac{p}{q} \right| < \frac{1}{q^{42}}$$

has only finitely many solutions.

Although the method of Thue is *ineffective*, i.e. one cannot bound the solutions of the equation

$$F(x, y) = k$$

it is possible to bound the number of solutions of this equation. Several mathematicians like Bombieri, Evertse, Győry, Schmidt, Siegel and others have contributed to this problem. See the recent book of Evertse and Győry [5] for various developments in this direction. We mention a result of Siegel [6].

A binary form $F(x, y) \in \mathbb{Z}[x, y]$ of degree r is said to be *diagonalisable* if it is of the form

$$(\alpha x + \beta y)^r - (\gamma x + \delta y)^r.$$

Here we consider forms with α, β, γ and δ algebraic satisfying $\alpha\delta - \beta\gamma \neq 0$ and $r \geq 3$. Let Δ be the discriminant of F. As $F(x, y) \in \mathbb{Z}[x, y]$, it is possible to write

$$(\alpha x + \beta y)(\gamma x + \delta y) = \chi(Ax^2 + Bxy + Cy^2), \ A, B, C \in \mathbb{Z}.$$

Let $D = D(F) = B^2 - 4AC$. Let $N_F(k)$ denote the number of primitive solutions (i.e. $\gcd(x, y) = 1$) of the inequality

$$|F(x, y)| \leq k.$$

Theorem 5.3.2 *Assume that $F(x, y)$ is a diagonalisable form of degree r and with discriminant Δ satisfying*

$$\Delta > 2^{r^2 - r} r^r k^{2r - 2} \left(r^4 k\right)^{c_l r^{2-l}}$$

where $r \geq 6 - l, l = 1, 2, 3,$

$$c_1 = 45 + \frac{593}{913}, \ c_2 = 6 + \frac{134}{4583} \ and \ c_3 = 75 + \frac{156}{167}.$$

Then

$$N_F(k) \leq \begin{cases} 2lr & \text{if } D < 0 \\ 4l & \text{if } D > 0, \ r \text{ is even and } F \text{ is indefinite} \\ 2l & \text{if } D > 0, \ r \text{ is odd and } F \text{ is indefinite} \\ 1 & \text{if } D > 0 \text{ and } F \text{ is definite.} \end{cases}$$

In particular, if $D < 0$ and $l = 1$, then $N_F(k) \leq 2r$ provided

$$|\Delta| > 2^{r^2 - r} r^{183.6r} k^{47.6r - 2}.$$

For a recent improvement of this result, see [7]. The most interesting families of diagonalisable forms are binomial forms, the forms of the shape $ax^r - by^r$. In an important work, Bennett [8] combined hyper-geometric method with Chebyshev like estimates for primes in arithmetic progressions to show that

$$ax^r - by^r = 1, a, b \in \mathbb{N}$$

has at most one solution in positive integers x and y. This is a best possible result since

$$(a + 1)x^r - ay^r = 1$$

has precisely one solution.

Although Dyson's improvement of Thue–Siegel theorem was mild, a lemma of Dyson proved to be very useful. Bombieri [9] generalised this lemma and used it to prove *effective* results on the approximation of certain algebraic numbers by rationals.

References

1. L.J. Mordell, *Diophantine Equations* (Academic, New York, 1969)
2. H. Davenport, A note on Thue's theorem. Mathematika **15**, 76–87 (1968)
3. C.S. Davis, Rational approximation to e. J. Aust. Math. Soc. **25**, 497–502 (1978)
4. K. Mahler, On the approximation π. Proc. Akad. Wetensch. Ser. A **56**, 30–42 (1953)
5. J.-H. Evertse, K. Győry, *Unit Equations in Diophantine Number Theory* (Cambridge University Press, Cambridge, 2015), 378 p.
6. C.L. Siegel, Einige Erläuterungen zu Thue's Unterschungen über Annäherungswerte algbraischer zahlen und diophantische. Gleichungen Nach Akad Wissen Göttingen Math-Phys, 169–195 (1970)
7. S. Akhtari, N. Saradha, D. Sharma, Thue's inequalities and hyper geometric method. Ramanujan J. **45**(2), 521–567 (2018)
8. M.A. Bennett, Rational approximation to algebraic numbers of small height: the Diophantine equation $|ax^n - by^n| = 1$,. J. Reine Angew. Math. **535**, 1–49 (2001)
9. E. Bombieri, On the Thue -Siegel -Dyson theorem. Acta Math. **148**, 255–296 (1982)

Chapter 6
Roth's Theorem

I'm not lost for I know where I am. But however, where I am may be lost.

–Winnie the Pooh

Thue's and Siegel's improvements of Liouville's theorem depend on the construction of an auxiliary polynomial in *two* variables possessing zeros to a high order. Any further progress seemed to require non-trivial extension of the arguments relating to polynomials in several variables especially the possible multiplicities of its zeros. This was discovered by Roth in 1955, when he proved that κ in Theorem 5.2.3 can be taken as $2 + \epsilon, \epsilon > 0$. To deal with the multiplicities of zeros of multi-variable polynomials, Roth introduced the notion of *index* of a polynomial; see Sect. 6.1. This notion was later used by Vojta in 1991 in his proof of Falting's famous theorem about Mordell conjecture. There are now other proofs of Roth's theorem available based on techniques of algebraic geometry. Roth was awarded the Fields Medal at ICM in Edinburgh in 1958.

We have already seen in Chap. 5 that Roth's theorem is essentially best possible with respect to the exponent $2 + \epsilon$. This can also be seen via continued fractions. Let h_n/k_n be the n-th convergent to the irrational number ξ. Then it is well known that

$$\left| \xi - \frac{h_n}{k_n} \right| < \frac{1}{k_n k_{n+1}} < \frac{1}{k_n^2}.$$

Thus there are infinitely many rational numbers h/k such that

$$\left| \xi - \frac{h}{k} \right| < \frac{1}{k^2}.$$

© Springer Nature Singapore Pte Ltd. 2020
S. Natarajan and R. Thangadurai, *Pillars of Transcendental Number Theory*,
https://doi.org/10.1007/978-981-15-4155-1_6

For proving the theorem, we follow Roth [1]. One may also refer to LeVeque [2] for approximation of ξ by algebraic numbers. We need some preparation.

6.1 Index of a Polynomial

We saw in the proof of Theorems of 5.2.3 and 5.3.1 that a polynomial in two variables was constructed. Roth used a polynomial $P(X_1, \ldots, X_m)$ in several variables. If $p_1/q_1, \ldots, p_m/q_m$ are very good rational approximations to α, then one may substitute these into $P(X_1, \ldots, X_m)$. The main difficulty is to show that $P(p_1/q_1, \ldots, p_m/q_m) \neq 0$. This was fairly easy in Lemmas 6.3.1 and 6.4.1 where $m = 2$. Roth used the polynomial described in Lemma 1.4.3 for this purpose. One also needs m very good rational approximations and it depends on κ.

A simple way to define the *order* of vanishing of $P(X_1, \ldots, X_m)$ at a given point $(\xi_1, \ldots, \xi_m)$ is to take the smallest value of $i_1 + \cdots + i_m$ for which the partial derivative

$$\left(\frac{\partial}{\partial X_1} \right)^{i_1} \cdots \left(\frac{\partial}{\partial X_m} \right)^{i_m} P \mid_{(\xi_1, \cdots, \xi_m)} := P^{(i_1, \ldots, i_m)}(\xi_1, \ldots, \xi_m) \neq 0. \qquad (6.1)$$

But it became necessary to study polynomials $P(X_1, \ldots, X_m)$ which have different degrees in $X_1, \ldots, X_m$ and hence it was better to attach different weights to $i_1, \ldots, i_m$.

Definition of Index
Let $P \in \mathbb{Z}[X_1, \ldots, X_m]$ and $\not\equiv 0$. The *index* of P at $(\xi_1, \ldots, \xi_m)$ with respect to a tuple $(r_1, \ldots, r_m)$ is defined to be the least value of

$$\frac{i_1}{r_1} + \cdots + \frac{i_m}{r_m}$$

for which (6.1) holds. If $P \equiv 0$, index is defined to be $+\infty$. We denote the index by $I_{P, r_1, \ldots, r_m}(\xi_1, \ldots, \xi_m)$. If $r_1, \ldots, r_m, \xi_1, \ldots, \xi_m$ remain the same, we omit referring to them and write simply I_P.

Note that $I_P \geq 0$ and $I_P = 0$ if and only if $P(\xi_1, \ldots, \xi_m) \neq 0$. Further, if

$$P^{(k_1, \ldots, k_m)}(X_1, \ldots, X_m) \not\equiv 0,$$

then its index at $(\xi_1, \ldots, \xi_m)$ is at least

$$I_P - \frac{k_1}{r_1} - \cdots - \frac{k_m}{r_m}.$$

We list here some properties which can be easily verified. Let $P(X_1, \ldots, X_m)$ and $Q(X_1, \ldots, X_m)$ be two non-identically vanishing polynomials. Let the indices be formed at $(\xi_1, \ldots, \xi_m)$. Then

$$I_{P+Q} \geq \min(I_P, I_Q) \text{ and } I_{PQ} = I_P + I_Q.$$

Suppose P is a polynomial in $X_1, \ldots, X_{m-1}$ and Q is a polynomial in X_m, then

$$I_{PQ,r_1,\ldots,r_m}(\xi_1, \ldots, \xi_m) = I_{P,r_1,\ldots,r_{m-1}}(\xi_1, \ldots, \xi_{m-1}) + I_{Q,r_m}(\xi_m).$$

We leave the proofs of these properties to the reader.

6.2 Set of Polynomials

In this section we define a set of polynomials in several variables and obtain an upper bound for the index of the polynomials in the set. Let us denote by $\mathcal{R}_m = \mathcal{R}_m(B; r_1, \ldots, r_m)$ the set of polynomials $R(X_1, \ldots, X_m)$ satisfying the following conditions.

1. $R(X_1, \ldots, X_m) \in \mathbb{Z}[X_1, \ldots, X_m]$ and $R \not\equiv 0$.

2. R is of degree at most r_j in X_j for $1 \leq j \leq m$.

3. $H(R) \leq B$.

Let $\psi_i = p_i/q_i$ with $q_i > 0$, $\gcd(p_i, q_i) = 1$ and $h(\psi_i) = q_i$ for $1 \leq i \leq m$. Let

$$\theta(R) = I_{R,r_1,\ldots,r_m}(\psi_1, \ldots, \psi_m).$$

Define

$$\Theta_m = \Theta_m(B; q_1, \ldots, q_m; r_1, \ldots, r_m) = \sup \theta(R)$$

where the supremum is taken over all $R \in \mathcal{R}_m$ and all rational numbers $\psi_1, \ldots, \psi_m$ of heights $q_1, \ldots, q_m$, respectively. Our aim is to bound Θ_m. We begin with the case $m = 1$.

Lemma 6.2.1 *We have*

$$\Theta_1(B; q_1; r_1) \leq \frac{\log B}{r_1 \log q_1}.$$

Proof Let $\theta = \theta(R)$. By the definition of index, $R(X_1)$ is divisible by $(X_1 - \psi_1)^{r_1 \theta}$ and since $R(X_1) \in \mathbb{Z}[X_1]$, $\gcd(p_1, q_1) = 1$, by Lemma 1.2.1, it is divisible by $(q_1 X_1 - p_1)^{r_1 \theta}$ and

$$R(X_1) = (q_1 X_1 - p_1)^{r_1\theta} Q(X_1)$$

for some $Q(X_1) \in \mathbb{Z}[X_1]$. Thus

$$q_1^{r_1\theta} \leq B$$

which gives the result. $\square$

Lemma 6.2.2 *Let $p \geq 2$ be a positive integer. Let $r_1, \ldots, r_p$ be positive integers such that*

$$r_p > \frac{10}{\delta}, \quad \frac{r_{j-1}}{r_j} > \frac{1}{\delta} \text{ for } 2 \leq j \leq p \tag{6.2}$$

where $0 < \delta < 1$ and $q_1, \ldots, q_p$ are positive integers. Further for any integer $1 \leq a \leq r_p + 1$, let

$$\Phi_a = \Theta_1(M; q_p; ar_p) + \Theta_{p-1}(M; q_1, \ldots, q_{p-1}; ar_1, \ldots, ar_{p-1}) \tag{6.3}$$

where

$$M = M_a = (r_1 + 1)^{pa} 2^{r_1 pa} a! B^a. \tag{6.4}$$

Then there exists an integer $1 \leq \ell \leq r_p + 1$ such that

$$\Theta_p(B; q_1, \ldots, q_p; r_1, \ldots, r_p) \leq 2(\Phi_\ell + \Phi_\ell^{1/2} + \delta^{1/2}). \tag{6.5}$$

Proof Let $R(X_1, \ldots, X_p)$ be any polynomial in the set $\mathcal{R}_p(B; r_1, \ldots, r_p)$ and let $\psi_i = p_i/q_i$ with $\gcd(p_i, q_i) = 1$ and $h(\psi_i) = q_i$ for $1 \leq i \leq p$. We prove that $\theta = \theta(R)$ satisfies (6.5).

By Lemma 1.4.3 and the Remark following it, there exists an integer ℓ and a polynomial $0 \not\equiv F(X_1, \ldots, X_p) \in \mathbb{Z}[X_1, \ldots, X_p]$ such that if

$$F(X_1, \ldots, X_p) = \det\left(\Delta_\mu \frac{1}{\nu!} \left(\frac{\partial}{\partial X_p}\right)^\nu R\right), 0 \leq \mu, \nu \leq \ell - 1,$$

then

$$F(X_1, \ldots, X_p) = U(X_1, \ldots, X_{p-1}) V(X_p)$$

where $U \in \mathbb{Z}[X_1, \ldots, X_{p-1}]$ with $\deg_{X_j} U \leq \ell r_j, 1 \leq j \leq p - 1$ and $V(X_p) \in \mathbb{Z}[X_p]$ with $\deg_{X_p} V \leq \ell r_p$. Further by (1.8),

$$H(F) \leq \{(r_1 + 1) \cdots (r_p + 1)\}^\ell 2^{\ell(r_1 + \cdots + r_p)} \ell! B^\ell.$$

An Upper Bound for the I_F

Since $r_1 > r_2 > \cdots > r_p$ by (6.2), we get

$$H(F) \leq M \text{ with } a = \ell.$$

Also

$$H(U) < M, H(V) < M.$$

The polynomial $U(X_1, \ldots, X_{p-1})$ is of degree at most ℓr_j in X_j for $1 \leq j \leq p - 1$. Hence

$$U \in \mathcal{R}_{p-1}(M; \ell r_1, \ldots, \ell r_{p-1}).$$

Thus its index at $(\psi_1, \ldots, \psi_{p-1})$ relative to $\ell r_1, \ldots, \ell r_{p-1}$ is at most

$$\Theta_{p-1}(M; q_1, \ldots, q_{p-1}; \ell r_1, \ldots, \ell r_{p-1}).$$

Hence it follows that its index at $(\psi_1, \ldots, \psi_{p-1})$ relative to $r_1, \ldots, r_{p-1}$ is at most

$$\ell \Theta_{p-1}(M; q_1, \ldots, q_{p-1}; \ell r_1, \ldots, \ell r_{p-1}).$$

Similarly $V(X_p) \in \mathcal{R}_1(M; \ell r_p)$ and its index at ψ_p relative to r_p is at most

$$\ell \Theta_1(M; q_p; \ell r_p).$$

Further by index property,
$$I_F = I_U + I_V \leq \ell \Phi_\ell. \tag{6.6}$$

A Lower Bound for I_F
The polynomial F is given by

$$F(X_1, \ldots, X_p) = \beta \det \left(\Delta_\mu \frac{1}{\nu!} \left(\frac{\partial}{\partial X_p} \right)^\nu R \right), 0 \leq \mu, \nu \leq \ell - 1.$$

Hence F is a sum of $\ell!$ terms and a typical term is of the form

$$\pm \beta(\Delta_{\mu_0} R) \left(\Delta_{\mu_1} \frac{1}{1!} \frac{\partial}{\partial X_p} R \right) \cdots \left(\Delta_{\mu_{\ell-1}} \frac{1}{(\ell-1)!} \left(\frac{\partial}{\partial X_p} \right)^{\ell-1} R \right), \tag{6.7}$$

where $\Delta_{\mu_0}, \ldots, \Delta_{\mu_{\ell-1}}$ are differential operators on $X_1, \ldots, X_{p-1}$ whose orders are at most $\ell - 1$. We shall determine a lower bound for the index of such a typical term. Let

$$\Delta = \frac{1}{i_1! \cdots i_{p-1}!} \left(\frac{\partial}{\partial X_1} \right)^{i_1} \cdots \left(\frac{\partial}{\partial X_{p-1}} \right)^{i_{p-1}}$$

of order $w = i_1 + \cdots + i_{p-1} \leq \ell - 1$. If

$$\Delta \frac{1}{\nu!} \left(\frac{\partial}{\partial X_p} \right)^\nu R(X_1, \ldots, X_p), \nu \le \ell - 1,$$

does not vanish identically, its index at $(\psi_1, \ldots, \psi_p)$ relative to $r_1, \ldots, r_p$ is at least

$$\theta - \frac{i_1}{r_1} - \cdots - \frac{i_{p-1}}{r_{p-1}} - \frac{\nu}{r_p} \ge \theta - \frac{w}{r_{p-1}} - \frac{\nu}{r_p}.$$

By (6.2) and $\ell \le r_p + 1$, we get

$$\frac{w}{r_{p-1}} \le \frac{\ell - 1}{r_{p-1}} \le \frac{r_p}{r_{p-1}} < \delta.$$

Thus the index of a term as in (6.7), which does not vanish identically, is at least

$$\sum_{\nu=0}^{\ell-1} \max \left(0, \theta - \frac{\nu}{r_p} \right) - \ell\delta.$$

Hence

$$I_F \ge \sum_{\nu=0}^{\ell-1} \max \left(0, \theta - \frac{\nu}{r_p} \right) - \ell\delta.$$

Now we shall complete the proof of the lemma.

Suppose $\theta r_p \le 10$. Then

$$\theta < \frac{10}{r_p} < \delta < 2\delta^{1/2}$$

and hence (6.5) is satisfied. Hence we shall assume that $\theta r_p > 10$. Then

$$[\theta r_p]^2 > 2\theta^2 r_p^2 / 3.$$

If now, $\theta r_p < \ell$, then we have

$$\sum_{\nu=0}^{\ell-1} \max \left(0, \theta - \frac{\nu}{r_p} \right) = r_p^{-1} \sum_{\nu=0}^{[\theta r_p]} (\theta r_p - \nu) \ge \frac{[\theta r_p]^2}{2r_p} \ge \frac{\theta^2 r_p}{3}.$$

If $\theta r_p \ge \ell$, then

$$\sum_{\nu=0}^{\ell-1} \max \left(0, \theta - \frac{\nu}{r_p} \right) = \sum_{\nu=0}^{\ell-1} \left(\theta - \frac{\nu}{r_p} \right) \ge \frac{\ell\theta}{2}.$$

Thus in either case,

$$I_F \geq \min\left(\frac{\ell\theta}{2}, \frac{r_p\theta^2}{3}\right) - \ell\delta. \tag{6.8}$$

We combine (6.6) and (6.8) to obtain

$$\min\left(\frac{\ell\theta}{2}, \frac{r_p\theta^2}{3}\right) \leq \ell(\Phi_\ell + \delta).$$

In the above inequality, suppose the minimum is $\frac{\ell\theta}{2}$, then $\theta < 2(\Phi_\ell + \delta)$ and hence (6.5) is satisfied. If the minimum is $\frac{r_p\theta^2}{3}$, then

$$\frac{r_p\theta^2}{3} \leq \ell(\Phi_\ell + \delta) \leq (r_p + 1)(\Phi_\ell + \delta) \leq \frac{4r_p(\Phi_\ell + \delta)}{3}.$$

Hence

$$\theta \leq 2(\Phi_\ell + \delta)^{1/2} \leq 2(\Phi_\ell^{1/2} + \delta^{1/2})$$

which completes the proof.   □

In the next lemma we impose some growth conditions on $h(\psi_1) = q_1, \ldots,$ $h(\psi_m) = q_m$ to obtain the following result.

Lemma 6.2.3 *Let m be a positive integer and assume that*

$$0 < \delta < \frac{1}{m}. \tag{6.9}$$

Let $r_1, \ldots, r_m$ be positive integers such that

$$r_m > \frac{10}{\delta}, \quad \frac{r_{j-1}}{r_j} > \frac{1}{\delta} \ for \ 2 \leq j \leq m. \tag{6.10}$$

Let $q_1, \ldots, q_m$ be positive integers such that

$$\log q_1 > m(2m + 1)/\delta, \tag{6.11}$$

and

$$r_j \log q_j \geq r_1 \log q_1 \ for \ 2 \leq j \leq m. \tag{6.12}$$

Then

$$\Theta_m(q_1^{\delta r_1}; q_1, \ldots, q_m; r_1, \ldots, r_m) < 10^m \delta^{1/2^m}.$$

Proof The proof is by induction on m. Let $m = 1$. By Lemma 6.2.1, (6.9) and (6.11), we get

$$\Theta_1(q_1^{\delta r_1}; q_1; r_1) < \frac{\log(q_1^{\delta r_1})}{r_1 \log q_1} = \delta \leq 10\delta^{1/2}$$

giving the desired inequality. Next suppose that $p \geq 2$ is an integer and the lemma holds for $m = p - 1$. We prove the result for $m = p$. By (6.9) and (6.10), Lemma 6.2.2 is valid. We shall estimate M.

Estimate for M
By (6.4)

$$M = (r_1 + 1)^{p\ell} 2^{r_1 p\ell} \ell! B^\ell \leq \left((r_1 + 1)^p 2^{r_1 p} \ell q_1^{\delta r_1} \right)^\ell$$

for some integer ℓ satisfying $\ell \leq r_p + 1 < r_1 + 1 \leq 2^{r_1}$. Hence

$$M < \left(2^{(2p+1)r_1} q_1^{\delta r_1} \right)^\ell < \left(e^{(2p+1)r_1} q_1^{\delta r_1} \right)^\ell.$$

By (6.11), with $m = p$ we have $2p + 1 < \delta \log q_1 / p$, so that

$$M < q_1^{\delta_1 \ell r_1} \text{ where } \delta_1 = \delta(1 + 1/p). \tag{6.13}$$

Note also that by (6.9) with $m = p$ we have

$$\delta_1 < \frac{(1 + 1/p)}{p} < \frac{1}{(p-1)}. \tag{6.14}$$

Estimate for Θ_1 and Θ_{p-1}
From (6.13) we get

$$\Theta_1(M; q_p; \ell r_p) \leq \Theta_1\left(q_1^{\delta_1 \ell r_1}; q_p; \ell r_p \right).$$

Note that by (6.10), we have

$$r_p < \delta^{p-1} r_1.$$

Using Lemma 6.2.1, (6.11) and (6.12) we therefore get

$$\Theta_1(M; q_p; \ell r_p) \leq \frac{\delta_1 \ell r_1 \log q_1}{\ell r_p \log q_p} \leq \delta_1 < \delta_1^{1/2}.$$

Further

$$\Theta_{p-1}(M; q_1, \ldots, q_{p-1}; \ell r_1, \ldots, \ell r_{p-1}) \leq \Theta_{p-1}\left(q_1^{\delta_1 \ell r_1}; q_1, \ldots, q_{p-1}; \ell r_1, \ldots, \ell r_{p-1} \right).$$

To estimate the right-hand side of the above inequality, we use induction hypothesis that the lemma holds when $m = p - 1$. The conditions of the lemma are satisfied for $m = p - 1$ by replacing δ by δ_1 with $\delta_1 > \delta$ satisfying (6.14) and replacing $r_1, \ldots, r_{p-1}$ by $\ell r_1, \ldots, \ell r_{p-1}$. Hence

$$\Theta_{p-1}\left(q_1^{\delta_1 \ell r_1}; q_1, \ldots, q_{p-1}; \ell r_1, \ldots, \ell r_{p-1}\right) < 10^{p-1} \delta_1^{1/2^{p-1}}.$$

Since $\delta_1 < 2\delta$ by (6.13), from the estimates for $\Theta_1(M; q_p; \ell r_p)$, $\Theta_{p-1}(M; q_1, \ldots, q_{p-1}; \ell r_1, \ldots, \ell r_{p-1})$ and (6.3), we get that

$$\Phi_\ell < 2\delta + 2\left(10^{p-1} \delta^{1/2^{p-1}}\right) < 3\left(10^{p-1} \delta^{1/2^{p-1}}\right).$$

Final Estimate for Θ_p

By (6.5), we get

$$\Theta_p\left(q_1^{\delta r_1}; q_1, \ldots, q_p; r_1, \ldots, r_p\right) < 2\left(3(10^{p-1} \delta^{1/2^{p-1}}) + 3^{1/2} 10^{(p-1)/2} \delta^{1/2^p} + \delta^{1/2}\right)$$

$$< 2\left(\frac{3}{10} + \frac{3^{1/2}}{10^{3/2}} + \frac{1}{10^2}\right) 10^p \delta^{1/2^p}$$

$$< 10^p \delta^{1/2^p}.$$

$\square$

6.3 A Combinatorial Lemma

Lemma 6.3.1 *Let $r_1, \ldots, r_m$ be positive integers and $\lambda > 0$. Let $S_m(\lambda)$ be the set of integers $(j_1, \ldots, j_m)$ satisfying the two conditions below.*

(i) $0 \le j_i \le r_i$ with $1 \le i \le m$,

(ii) $\dfrac{j_1}{r_1} + \cdots + \dfrac{j_m}{r_m} \le \dfrac{1}{2}(m - \lambda).$

Then

$$|S_m(\lambda)| \le 2m^{1/2} \lambda^{-1}(r_1 + 1) \cdots (r_m + 1).$$

Proof The proof is by induction on m. Suppose $m = 1$. Then the number of integers j_1 satisfying $0 \le j_1 \le r_1$ and $j_1 \le \frac{1}{2}(1 - \lambda)r_1$ is at most $r_1 + 1$ if $\lambda \le 1$ and is 0 if $\lambda > 1$ and hence the result is true for $m = 1$.

Let $m > 1$. Suppose $\lambda \le 2m^{1/2}$. Then

$$2m^{1/2} \lambda^{-1}(r_1 + 1) \cdots (r_m + 1) \ge (r_1 + 1) \cdots (r_m + 1).$$

The product on the right-hand side is the total number of m-tuples satisfying (i). Hence the result is true in this case.

Now we assume that $\lambda > 2m^{1/2}$. Fix j_m. We count the number of $(m - 1)$-tuples $(j_1, \ldots, j_{m-1})$ such that

(i)' $0 \le j_i \le r_i$ with $1 \le i \le m - 1$,

$(ii)'$ $\dfrac{j_1}{r_1} + \cdots + \dfrac{j_{m-1}}{r_{m-1}} \le \dfrac{1}{2}\left(m - \lambda - \dfrac{2j_m}{r_m}\right).$

Put

$$\lambda' = \lambda'(j_m) = \lambda - 1 + \frac{2j_m}{r_m}.$$

Then

$$|S_m(\lambda)| = \sum_{j_m=0}^{r_m} |S_{m-1}(\lambda'(j_m))|.$$

Hence by induction hypothesis,

$$|S_m(\lambda)| \le 2(m-1)^{1/2}(r_1+1)\cdots(r_{m-1}+1)\sum_{j=0}^{r_m}\left(\lambda - 1 + \frac{2j}{r_m}\right)^{-1}.$$

Thus it is enough to show that

$$\sum_{j=0}^{r}\left(\lambda - 1 + \frac{2j}{r}\right)^{-1} \le \lambda^{-1}(m-1)^{-1/2}m^{1/2}(r+1)$$

for all positive integers r and m.

Let r be even. Substitute $j = \frac{r}{2} + k$. Then

$$\sum_{j=0}^{r}\left(\lambda - 1 + \frac{2j}{r}\right)^{-1} = \sum_{k=-r/2}^{r/2}\left(\lambda + \frac{2k}{r}\right)^{-1}$$

$$= \lambda^{-1} + \sum_{k=1}^{r/2}\left\{\left(\lambda + \frac{2k}{r}\right)^{-1} + \left(\lambda - \frac{2k}{r}\right)^{-1}\right\}$$

$$= \lambda^{-1} + \sum_{k=1}^{r/2} 2\lambda\left(\lambda^2 - \frac{4k^2}{r^2}\right)^{-1}$$

$$\le \lambda^{-1} + 2\lambda\sum_{k=1}^{r/2}(\lambda^2 - 1)^{-1} = \lambda^{-1} + 2\lambda^{-1}\sum_{k=1}^{r/2}(1 - \lambda^{-2})^{-1}$$

$$\le (r+1)\lambda^{-1}(1 - \lambda^{-2})^{-1}.$$

By our assumption on λ, $1 - \lambda^{-2} > 1 - (1/4m) > (1 - 1/m)^{1/2}$. Hence we get the desired inequality.

Let now r be odd. Put $j = (r-1)/2 + k$. Then

$$\sum_{j=0}^{r}\left(\lambda-1+\frac{2j}{r}\right)^{-1} = \sum_{k=-(r-1)/2}^{(r+1)/2}\left(\lambda+\frac{2k-1}{r}\right)^{-1}$$

$$= \sum_{k=1}^{(r+1)/2}\left(\left(\lambda+\frac{2k-1}{r}\right)^{-1}+\left(\lambda-\frac{2k-1}{r}\right)^{-1}\right)$$

$$= 2\lambda\sum_{k=1}^{(r+1)/2}\left(\lambda^2-\frac{(2k-1)^2}{r^2}\right)^{-1}$$

$$\le (r+1)\lambda(\lambda^2-1)^{-1} = (r+1)\lambda^{-1}(1-\lambda^{-2})^{-1}.$$

and the result follows as before. $\square$

6.4 The Approximation Polynomial

Let α be an algebraic integer of degree $n \ge 2$ with

$$h(\alpha) = A.$$

Note that $\lceil\alpha\rceil \le A + 1$ (see Chap. 1, Exercise 2).

Choice of Parameters
We choose the values $m, \delta, q_1, \ldots, q_m, r_1, \ldots, r_m$ satisfying the following conditions.

P1. $0 < \delta < \dfrac{1}{m}$

P2. $10^m \delta^{1/2^m} + 2n(1+3\delta)m^{1/2} < \dfrac{m}{2}$

P3. $r_m > \dfrac{10}{\delta}, \dfrac{r_{j-1}}{r_j} > \dfrac{1}{\delta}$ for $2 \le j \le m$

P4. $\delta^2 \log q_1 > 2m + 1 + 4m \log(A + 1)$

P5. $r_j \log q_j \ge r_1 \log q_1$ for $2 \le j \le m$.

From P4 and P1, we see that

$$\log q_1 > (2m + 1)/\delta^2 > m(2m + 1)/\delta.$$

Thus (6.11) is valid. The conditions (6.9), (6.10) and (6.12) are listed as conditions P1,P3 and P5, respectively. Thus Lemma 6.2.3 is valid. We put

(i) $\lambda = 4n(1 + 3\delta)m^{1/2}$;

(ii) $\mu = \dfrac{m - \lambda}{2}$;

(iii) $\eta = 10^m \delta^{1/2^m}$;

(iv) $B_1 = [q_1^{\delta r_1}]$.

Using these notations, we see from the condition P2 above that

$$\eta < \mu; \ q_1^{\delta r_1/2} < B_1.$$

Further by conditions P4 and P1, we also have

$$B_1 > 4; \ B_1^\delta > 2^{mr_1}; \ B_1^\delta > (A+1)^{2mr_1}. \tag{6.15}$$

The following lemma will be the *main lemma* to be used in the proof of Roth's theorem. For any polynomial $P(X_1, \ldots, X_m) \in \mathbb{Z}[X_1, \ldots, X_m]$, we put

$$P_{i_1, \ldots, i_m} = P_{i_1, \ldots, i_m}(X_1, \ldots, X_m)$$

$$= \frac{1}{i_1! \cdots i_m!} \left(\frac{\partial}{\partial X_1} \right)^{i_1} \cdots \left(\frac{\partial}{\partial X_m} \right)^{i_m} P$$

with integers $i_1 \geq 0, \ldots, i_m \geq 0$.

Lemma 6.4.1 *Assume that conditions P1–P5 on the choice of the parameters are satisfied. Suppose that $\psi_i = p_i/q_i$ with $\gcd(p_i, q_i) = 1$, $h(\psi_i) = q_i$ for $1 \leq i \leq m$. Then there exists $Q(X_1, \ldots, X_m) \in \mathbb{Z}[X_1, \ldots, X_m]$ of degree at most r_j in X_j for $1 \leq j \leq m$ having the following properties.*

(a) $I_{Q, r_1, \ldots, r_m}(\psi_1, \ldots, \psi_m) \geq \mu - \eta.$

(b) $Q(\psi_1, \ldots, \psi_m) \neq 0.$

(c) $|Q_{i_1, \ldots, i_m}(X_1, \ldots, X_m)| < B_1^{1+3\delta}(1 + |X_1|)^{r_1} \cdots (1 + |X_m|)^{r_m}.$

Proof Estimation for $\boxed{P_{j_1, \ldots, j_m}(\alpha, \ldots, \alpha)}$.
    Consider

$$P(X_1, \ldots, X_m) = \sum_{s_1=0}^{r_1} \cdots \sum_{s_m=0}^{r_m} \gamma(s_1, \ldots, s_m) X_1^{s_1} \cdots X_m^{s_m} \in \mathbb{Z}[X_1, \ldots, X_m]$$

with $0 \leq \gamma(s_1, \ldots, s_m) \leq B_1$. Let

$$(r_1 + 1) \cdots (r_m + 1) = r.$$

Then there are $(B_1 + 1)^r$ distinct polynomials $P(X_1, \ldots, X_m)$. Note that

$$P_{j_1, \ldots, j_m}(X_1, \ldots, X_m) = \sum_{s_1=j_1}^{r_1} \cdots \sum_{s_m=j_m}^{r_m} \gamma(s_1, \ldots, s_m) \binom{s_1}{j_1} \cdots \binom{s_m}{j_m} X_1^{s_1-j_1} \cdots X_m^{s_m-j_m}.$$

Then by (6.15), we have

$$H(P_{j_1,\ldots,j_m}) \le 2^{r_1+\cdots+r_m} B_1 \le 2^{mr_1} B_1 < B_1^{1+\delta}.$$

Also

$$A^{r_1+\cdots+r_m} \le (A+1)^{mr_1} < B_1^{\delta}.$$

Hence

$$\left| P_{j_1,\ldots,j_m}(\alpha,\ldots,\alpha) \right| \le (r_1+1)\cdots(r_m+1) B_1^{1+\delta} A^{r_1+\cdots+r_m} \le (2A)^{mr_1} B_1^{1+\delta} \le B_1^{1+3\delta}.$$

Application of Pigeonhole Principle

Let $L = \mathbb{Q}(\alpha)$ and order the conjugates of α as follows. Let $\alpha_1,\ldots,\alpha_{\rho_1}$ be real and $\alpha_{\rho_1+\nu}$ and $\alpha_{\rho_1+\rho_2+\nu}$ be complex conjugates for $1 \le \nu \le \rho_2$ so that $\rho_1 + 2\rho_2 = n$. Let us fix

$$\phi = P_{j_1,\ldots,j_m}(\alpha,\ldots,\alpha)$$

for some $(j_1,\ldots,j_m)$ satisfying

$$0 \le j_1 \le r_1,\ldots,0 \le j_m \le r_m, \quad \frac{j_1}{r_1}+\cdots+\frac{j_m}{r_m} \le \mu. \tag{6.16}$$

Then ϕ is a polynomial in α with rational coefficients and let its conjugates be denoted by $\phi^{(\nu)}$, $1 \le \nu \le n$. Define n real numbers by the equations

$$\phi_\nu = \phi^{(\nu)}, 1 \le \nu \le \rho_1,$$

$$\phi_\nu + i\phi_{\nu+\rho_2} = \phi^{(\nu)}, \rho_1 + 1 \le \nu \le \rho_1 + \rho_2.$$

Let us consider the tuple $(\phi_1,\ldots,\phi_n)$ for $(j_1,\ldots,j_m)$ satisfying (6.16). By Lemma 6.3.1, there are

$$M \le 2nm^{1/2}r/\lambda$$

tuples, and each element of the tuple has its absolute value bounded by $\lceil B_1^{1+3\delta} \rceil = t$. Hence all the tuples lie in a cube of edge $2t$ in an M-dimensional space. Dividing each edge into $3t$ equal parts we get $(3t)^M$ subcubes of edge $2/3$. There are $(B_1 + 1)^r$ distinct polynomials. As $B_1 > 4$ by (6.15), we have

$$(B_1 + 1)^r > (3t)^M.$$

Hence points corresponding to two different polynomials P^* and P^{**} in $X_1,\ldots,X_m$ variables lie in the same cube and taking

$$\overline{P}(X_1,\ldots,X_m) = P^*(X_1,\ldots,X_m) - P^{**}(X_1,\ldots,X_m),$$

we get

$$\left| \overline{P}_{j_1,\ldots,j_m}(\alpha,\ldots,\alpha) \right| \leq \sqrt{2} \times 2/3 < 1$$

for some $(j_1,\ldots,j_m)$ satisfying (6.16). Since $\overline{P}_{j_1,\ldots,j_m}(\alpha,\ldots,\alpha)$ is an algebraic integer, it must be zero. Hence

$$I_{\overline{P}}(\alpha,\ldots,\alpha) \geq \mu.$$

Further the coefficients of $\overline{P}$ are all not zero and $H(\overline{P}) \leq B_1$.

Application of Lemma 6.2.3

Note that $\overline{P} \in \mathcal{R}_m(q_1^{\delta r_1}; r_1,\ldots,r_m)$ since P^* and P^{**} have integer coefficients in $[0, B_1]$. Hence by Lemma 6.2.3, we have

$$I_{\overline{P},r_1,\ldots,r_m}(\psi_1,\ldots,\psi_m) < \eta.$$

Therefore there exists $Q(X_1,\ldots,X_m)$ given by

$$Q(X_1,\ldots,X_m) = \frac{1}{k_1!\cdots k_m!}\left(\frac{\partial}{\partial X_1}\right)^{k_1}\cdots\left(\frac{\partial}{\partial X_m}\right)^{k_m}\overline{P},$$

with

$$\frac{k_1}{r_1} + \cdots + \frac{k_m}{r_m} < \eta$$

such that $Q(\psi_1,\ldots,\psi_m) \neq 0$. Also

$$I_{Q,r_1,\ldots,r_m}(\alpha,\ldots,\alpha) \geq \mu - \eta.$$

Further

$$H(Q) \leq 2^{r_1+\cdots+r_m} B_1 < B_1^{1+\delta}.$$

Hence for arbitrary derivative of Q we have

$$H(Q_{i_1,\ldots,i_m}) \leq 2^{r_1+\cdots+r_m} B_1^{1+\delta} < B_1^{1+2\delta}.$$

Finally,

$$|Q_{i_1,\ldots,i_m}(X_1,\ldots,X_m)| < B_1^{1+3\delta}\prod_{\nu=1}^{m}(1+|X_\nu|+\cdots+|X_\nu|^{r_\nu}) < B_1^{1+3\delta}\prod_{\nu=1}^{m}(1+|X_\nu|)^{r_\nu}.$$

Similar inequalities hold for conjugate polynomials as well. $\square$

6.5 Statement and Proof of Roth's Theorem

Theorem 6.5.1 *Suppose α is an algebraic number of degree $n \geq 2$. Then for each $\kappa > 2$ the inequality*

$$\left| \alpha - \frac{p}{q} \right| < \frac{1}{q^{\kappa}} \tag{6.17}$$

has only finitely many solutions in p/q.

Proof We may take α as an algebraic integer. From now on we assume that (6.17) holds for infinitely many reduced rationals p/q with $\kappa = 2 + \epsilon$ for some $\epsilon > 0$. See (a)–(e) in Sect. 6.1 for justifying these assumptions. □

Choice of m and δ

Take $\kappa = 2 + \epsilon$ with $\epsilon > 0$. Choose m such that

$$m > 16n^2 \left(\frac{\kappa}{\kappa - 2} \right)^2.$$

Thus $m > 4nm^{1/2}$ and

$$\kappa > \frac{2m}{m - 4nm^{1/2}}.$$

Note that $\eta = 10^m \delta^{1/2^m}$ becomes arbitrarily small with δ. Hence for sufficiently small δ

$$m - 4n(1 + 3\delta)m^{1/2} - 2\eta > 0.$$

This is same as condition P2 in Sect. 6.5. We also choose δ so that P1 is satisfied and

$$\kappa(m - 4nm^{1/2}) - 2m \geq \kappa(12\delta nm^{1/2} + 2\eta) + (2m + 4 + 10\delta)\delta.$$

This gives

$$\kappa > \frac{2m(1 + \delta) + 2\delta(2 + 5\delta)}{m - 4(1 + 3\delta)nm^{1/2} - 2\eta}$$

which is equivalent to

$$\kappa > \frac{m(1 + \delta) + \delta(2 + 5\delta)}{\mu - \eta}. \tag{6.18}$$

Choice of $q_1, \ldots, q_m$

First choose a solution p_1/q_1 of (6.17) with q_1 large so that condition P4 is satisfied. Then choose $p_2/q_2, \ldots, p_m/q_m$ such that

$$\frac{\log q_j}{\log q_{j-1}} > \frac{2}{\delta}, \quad 2 \leq j \leq m.$$

Choice of $r_1, \ldots, r_m$

Take r_1 to be any integer such that

$$r_1 > \frac{10 \log q_m}{\delta \log q_1} \qquad (6.19)$$

and define r_j by

$$\frac{r_1 \log q_1}{\log q_j} \le r_j < \frac{r_1 \log q_1}{\log q_j} + 1, 2 \le j \le m. \qquad (6.20)$$

Then condition P5 is satisfied. Also

$$\frac{r_j \log q_j}{r_1 \log q_1} < 1 + \frac{\log q_j}{r_1 \log q_1} \le 1 + \frac{\log q_m}{r_1 \log q_1} < 1 + \frac{\delta}{10}. \qquad (6.21)$$

By (6.20) and (6.19), we get

$$r_m \ge \frac{r_1 \log q_1}{\log q_m} > \frac{10}{\delta}.$$

By (6.20) and (6.21), we get

$$\frac{r_{j-1}}{r_j} > \frac{\log q_j}{\log q_{j-1}} \left(1 + \frac{\delta}{10}\right)^{-1} > \delta^{-1}.$$

Hence P3 is also satisfied.

By the above choices of $m, \delta, q_1, \ldots, q_m, r_1, \ldots, r_m$, all the conditions P1–P5 are satisfied. So Lemma 6.4.1 holds.

Application of Lemma 6.4.1

By Lemma 6.4.1, there exists a polynomial $Q(X_1, \ldots, X_m)$ satisfying the properties (a)–(c) listed therein. Thus for a set of m reduced rationals $p_1/q_1, \ldots, p_m/q_m$, the number

$$\phi_0 = Q\left(\frac{p_1}{q_1}, \ldots, \frac{p_m}{q_m}\right) \ne 0 \text{ and is in } \mathbb{Q}$$

and hence there exist $k_1, \ldots, k_m \in \mathbb{N}$ such that

$$|k_1^{r_1} \cdots k_m^{r_m} \phi_0| \ge 1.$$

On the other hand, we have

$$Q\left(\frac{p_1}{q_1}, \ldots, \frac{p_m}{q_m}\right) = \sum_{i_1=0}^{r_1} \cdots \sum_{i_m=0}^{r_m} Q_{i_1,\ldots,i_m}(\alpha, \ldots, \alpha) \left(\frac{p_1}{q_1} - \alpha\right)^{i_1} \cdots \left(\frac{p_m}{q_m} - \alpha\right)^{i_m}$$

and the terms with

$$\frac{i_1}{r_1} + \cdots + \frac{i_m}{r_m} < \mu - \eta$$

vanish. In the other terms we have

$$\left|\left(\frac{p_1}{q_1} - \alpha\right)^{i_1} \cdots \left(\frac{p_m}{q_m} - \alpha\right)^{i_m}\right| < (q_1^{i_1} \cdots q_m^{i_m})^{-\kappa}$$
$$= (q_1^{i_1/r_1}(q_2^{r_2/r_1})^{i_2/r_2} \cdots (q_m^{r_m/r_1})^{i_m/r_m})^{-r_1\kappa}$$
$$\le (q_1^{i_1/r_1} \cdots q_1^{i_m/r_m})^{-r_1\kappa} < q_1^{-r_1(\mu-\eta)\kappa}$$

since $q_j^{r_j/r_1} > q_1$ by (6.20). Hence by Lemma 6.4.1(c) and (6.15), we get

$$|\phi_0| < (r_1 + 1) \cdots (r_m + 1)B_1^{1+3\delta}(A + 1)^{mr_1}q_1^{-r_1(\mu-\eta)\kappa}$$
$$< B_1^{1+5\delta}q_1^{-r_1(\mu-\eta)\kappa}.$$

Thus

$$1 \le |k_1^{r_1} \cdots k_m^{r_m}\phi_0| \le B_1^{1+5\delta}q_1^{-r_1(\mu-\eta)\kappa}\prod_{i=1}^{m}q_i^{r_i}$$
$$< q_1^{\delta r_1(1+5\delta)+r_1+\cdots+r_m-r_1(\mu-\eta)\kappa}$$
$$< q_1^{\delta r_1(1+5\delta)+r_1m(1+\delta)-r_1(\mu-\eta)\kappa}$$

since $r_j < \delta r_{j-1}$ for $2 \le j \le m$. Hence

$$\delta(1 + 5\delta) + m(1 + \delta) > (\mu - \eta)\kappa$$

which gives

$$\kappa < \frac{m(1 + \delta) + \delta(1 + 5\delta)}{\mu - \eta}$$

a contradiction to (6.18). $\square$

Exercise

(1) Show that

$$\sum_{n=0}^{\infty} 2^{-[\theta^n]}$$

is transcendental if $\theta > 2$. Here $[x]$ denotes the greatest integer in x.

(2) Let $k \in \mathbb{N}$. Suppose $3^k = 2^k q + r$, then show that $0 < r < 2^k - q$ for k sufficiently large. (Hint: Use Theorem 6.5.2 below).

Notes

In 1957, Ridout [3] obtained an extension of Roth's theorem which is as follows.

Theorem 6.5.2 *Let α be a non-zero algebraic number and let $p_1, \ldots, p_r, q_1, \ldots, q_s$ be distinct prime numbers. Let μ, ν and c be real numbers with $0 \le \mu, \nu \le 1$ and $c > 0$. Let p and q be integers of the form*

$$p = p^* p_1^{a_1} \cdots p_r^{a_r}, q = q^* q_1^{b_1} \cdots q_s^{b_s}$$

with a_i's and b_j's non-negative integers and p^* and q^* are non-zero integers satisfying

$$|p^*| \le cp^\mu; \ |q^*| \le cq^\nu.$$

Further suppose that $\kappa > \mu + \nu$. Then there are only a finite number of solutions to (5.5).

We recover Roth's theorem by taking $\mu = \nu = c = 1$. Roth's theorem and Ridout's theorem can be applied to establish that α is transcendental if it admits infinitely many very good rational approximants. For instance, the transcendence of the Champernowne number $0.1234567891011\ldots$, can be proven using these theorems. For this and other results, see [4] and references given therein.

In Roth's theorem, it is believable that the factor q^ϵ could be replaced by a smaller factor. By Theorem 5.1.3 of Khintchine, we know that for almost all $\alpha \in \mathbb{R}$

$$\left| \alpha - \frac{p}{q} \right| < \frac{1}{q^2 (\log q)^{1+\epsilon}}$$

has only finitely many solutions for every $\epsilon > 0$. It was conjectured by Lang [5] that the same conclusion holds to be true for all algebraic irrationals. In 1959, Cugiani [6] could show that
if the rational numbers $\frac{p(1)}{q(1)}, \frac{p(2)}{q(2)}, \ldots$ are solutions of

$$\left| \alpha - \frac{p}{q} \right| < \frac{1}{q^{2+20(\log \log \log q)^{-1/2}}}$$

with $0 < q(1) < q(2) < \cdots$, then

$$\limsup \frac{\log q(k+1)}{\log q(k)} = \infty.$$

Similar results were proved earlier by Siegel and Schneider .

For analogous results of Roth's theorem over function fields, see the article of Thakur [7].

References

1. K.F. Roth, Rational approximations to algebraic numbers. Mathematika **2**, 1–20 (1955). Corrigendum **2**, 168 (1955)
2. W.J. LeVeque, *Topics in Number Theory, Vol I and II* (Dover Publication Inc, New York, 1984)
3. D. Ridout, Rational approximations to algebraic numbers. Mathematika **4**, 125–131 (1957)
4. Y. Bugeaud, *Approximation by Algebraic numbers* (Cambridge University Press, Cambridge, 2004), 274 pp
5. S. Lang, Report on diophantine approximations. Bull. de la Soc. Math. de France **93**, 117–192 (1965)
6. M. Cugiani, Sulla approssimabilitá dei numeri algebrici mediante numeri razionali. Ann. Mat. Pura Appl. (4) **48**, 135–145 (1959)
7. D.S. Thakur, *Diophantine Approximation and Transcendence in Finite Characteristic*. Diophantine Equations, ed. by N. Saradha (Narosa Publishing House, New Delhi, 2005), pp. 265–278

Chapter 7
Baker's Theorems and Applications

The cave you fear to enter holds the treasure you seek

–Joseph Campbell

Gelfond and Schneider's theorem can be restated as follows.

For any algebraic number $\alpha \neq 0$, 1 the number $\log \alpha$ to any algebraic base other than 0 or 1 is either rational or transcendental.

Gelfond, by a refinement of his method, obtained a positive lower bound for the absolute value of $\beta_1 \log \alpha_1 + \beta_2 \log \alpha_2$ where β_1, β_2 denote algebraic numbers not both 0, and α_1, α_2 denote algebraic numbers not 0 or 1, with $\log \alpha_1 / \log \alpha_2$ irrational. Gelfond also remarked that an analogous theorem for linear forms in arbitrarily many logarithms of algebraic numbers would be of great value for the solution of some apparently very difficult problems in number theory. In 1966–68, Baker established such a result. See his papers [1] and his prize winning book [2]. Corollaries 7.2.1, 7.2.2 and 7.2.3 resolve the multidimensional analogue of Hilbert's seventh problem. We have chosen to give as applications, some results on Pillai's equation, the growth of the greatest prime factor of polynomial values and *effective* version of Thue' theorem; see Sects. 7.3 and 7.4.

At the ICM in Nice in 1970, Baker was awarded Fields Medal for his contributions to linear forms in the logarithms of algebraic numbers and their applications to various problems in number theory.

© Springer Nature Singapore Pte Ltd. 2020
S. Natarajan and R. Thangadurai, *Pillars of Transcendental Number Theory*,
https://doi.org/10.1007/978-981-15-4155-1_7

7.1   Statement of Baker's Theorems

Define the complex logarithm by $\log z = \log |z| + i \arg z$ with $-\pi < \arg z \leq \pi$. In fact one may take any branch of the logarithm. We denote by $c_{7.m} = c_{7.m}(\cdots), m \geq 1$, effectively computable positive numbers and we will specify within brackets the parameters on which the number depends.

Theorem 7.1.1 *Let* $\alpha_1, \ldots, \alpha_n \in \mathbb{A}\backslash\{0, 1\}$, $\beta_0 \in \mathbb{A}$ *and* $\beta_1, \ldots \beta_n \in \mathbb{A}\backslash\{0\}$. *Assume that*

$$\log \alpha_1, \ldots, \log \alpha_n \text{ are linearly independent over } \mathbb{Q}.$$

Then $\Lambda := \beta_0 + \beta_1 \log \alpha_1 + \cdots + \beta_n \log \alpha_n \neq 0$. *In other words,*

$$1, \log \alpha_1, \ldots, \log \alpha_n$$

are linearly independent over $\mathbb{A}$.

Once non-vanishing of Λ is known, one would like to find a non-trivial lower bound for $|\Lambda|$. Such a bound proved to have several applications in Diophantine equations. We shall present some such results and give a few applications. The following is a quantitative result of the above theorem proved by Baker himself in 1975.

Theorem 7.1.2 *Let* $\alpha_1, \ldots, \alpha_n \in \mathbb{A}\backslash\{0, 1\}$ *and* $\beta_0, \beta_1, \ldots \beta_n \in \mathbb{A}$. *Assume that* $\Lambda \neq 0$. *Then*

$$|\Lambda| \geq (eB)^{-c_{7.1}}$$

where $B = \max(h(\beta_0), h(\beta_1), \ldots, h(\beta_n))$ *and* $c_{7.1} = c_{7.1}(n, \alpha_1, \ldots, \alpha_n)$.

In 1977, Baker proved a more explicit bound as follows.

Theorem 7.1.3 *Let* $\alpha_1, \ldots, \alpha_n \in \mathbb{A}\backslash\{0, 1\}$. *Let the field* $\mathbb{Q}(\alpha_1, \ldots, \alpha_n)$ *have degree at most* d *over* $\mathbb{Q}$. *Let* $h(\alpha_j) \leq A_j$ *with* $A_j \geq 4$ *for* $1 \leq j \leq n$. *Put*

$$A = \max(A_j), \ \Omega = (\log A_1) \cdots (\log A_n), \ \Omega' = (\log A_1) \cdots (\log A_{n-1}).$$

Let $\beta_0, \beta_1, \ldots \beta_n \in \mathbb{Z}$ *with* $e^B \geq 4$. *Assume that* $\Lambda \neq 0$. *Then*

$$\log |\Lambda| \geq -(16nd)^{200n} (\log B) \Omega \log \Omega'.$$

Although completely explicit, the above bound has some drawbacks. The factor $(16nd)^{200n}$ is very large. It is expected that it can be replaced by a polynomial expression in n and d, and the constants can be greatly reduced. The product of the logarithms in Ω was expected to be replaced by the sum of the logarithms. After the efforts of several authors, Baker and Wüstholz [3] proved the following improved result in 1993.

Theorem 7.1.4 *Under the conditions of Theorem 7.1.3, we have either* $\Lambda = 0$ *or*

$$\log |\Lambda| \geq -(16nd)^{2(n+2)} (\log B)\Omega.$$

For many applications, only two or three logarithms occur. Best results in these cases were obtained by Laurent, Mignotte and Nesterenko and Bennett et al.; see [4] and [5]. Lastly, we state the result of Matveev [6] in 2000 which is the best known so far.

Theorem 7.1.5 *Let* $\alpha_1, \ldots, \alpha_n \in \mathbb{A}\backslash\{0, 1\}$. *Let* $h^\circ(\alpha_j), 1 \leq j \leq n$ *denote the absolute logarithmic height. Let* $\mathcal{K} = \mathbb{Q}(\alpha_1, \ldots, \alpha_n)$ *have degree at most d over* $\mathbb{Q}$. *Let*

$$\kappa = \begin{cases} 1 & \text{if } \mathcal{K} \subset \mathbb{R} \\ 2 & \text{if } \mathcal{K} \subset \mathbb{C}. \end{cases}$$

Consider

$$\Lambda = b_1 \log \alpha_1 + \cdots + b_n \log \alpha_n \text{ with } b_j \in \mathbb{Z} \text{ for } 1 \leq j \leq n.$$

Put

$$B = \max(|b_1|, \ldots, |b_n|)$$

and let A_j *be real numbers such that*

$$A_j \geq \max(dh^\circ(\alpha_j), |\log \alpha_j|, 0.16) \text{ for } 1 \leq j \leq n.$$

Then either $\Lambda = 0$ *or*

$$\log |\Lambda| \geq -c_{7.2}d^2 A_1 \cdots A_n \log(ed) \log(eB)$$

where

$$c_{7.2} = \min(\kappa^{-1}(en/2)^\kappa 30^{n+3} n^{3.5}, 2^{6n+20}).$$

There are innumerable papers in the literature giving various applications of the above effective results. We shall give few results here. The reader may see the books of Bugeaud [7], Shorey and Tijdeman [8] and the references therein for many other results.

7.2 Applications of the Qualitative Result—Theorem 7.1.1

Theorem 7.1.1 with $n = 1$ is Hermite–Lindemann–Weierstrass theorem and $n = 2$ is Gelfond–Schneider theorem. We derive some more results in the following corollaries.

Corollary 7.2.1 *Let* $\alpha_1, \ldots, \alpha_n \in \mathbb{A}\backslash\{0\}$ *and* $\beta_1, \ldots \beta_n \in \mathbb{A}$. *Then*

$$\beta_1 \log \alpha_1 + \cdots + \beta_n \log \alpha_n$$

is either zero or transcendental.

Proof When $n = 1$, the statement is Hermite–Lindemann–Weierstrass theorem. We prove the corollary by induction on n. We assume that the corollary is true for any $m < n$. We prove the corollary for $n = m$. Suppose the assertion does not hold for $n = m$. Then there exist $\beta_0, \beta_1, \ldots, \beta_m \in \mathbb{A}$ such that

$$\beta_1 \log \alpha_1 + \cdots + \beta_m \log \alpha_m = \beta_0 \qquad (7.1)$$

with $\beta_0 \neq 0$. Then by Theorem 7.1.1, $\log \alpha_1, \ldots, \log \alpha_m$ are linearly dependent over $\mathbb{Q}$. Hence there exist rational numbers $c_1, \ldots, c_m$, not all zero, such that

$$c_1 \log \alpha_1 + \cdots + c_m \log \alpha_m = 0.$$

We may assume without loss of generality that $c_m \neq 0$. Using this equation along with (7.1), we eliminate $\log \alpha_m$ to get a linear form

$$\beta_1' \log \alpha_1 + \cdots + \beta_{m-1}' \log \alpha_{m-1} = c_m \beta_0$$

where $c_m \beta_0 \neq 0$ and algebraic. By induction, the left-hand side is transcendental which is a contradiction. This proves the corollary.   □

As a consequence of Corollary 7.2.1 we show the following result.

Corollary 7.2.2 *The number*

$$e^{\beta_0} \alpha_1^{\beta_1} \cdots \alpha_n^{\beta_n}$$

is transcendental for $\alpha_1, \ldots, \alpha_n, \beta_0, \beta_1, \ldots \beta_n \in \mathbb{A}\setminus\{0\}$.

Proof Suppose $e^{\beta_0} \alpha_1^{\beta_1} \cdots \alpha_n^{\beta_n}$ equals an algebraic number α_{n+1}, then $\alpha_{n+1} \neq 0$ and

$$\beta_1 \log \alpha_1 + \cdots + \beta_n \log \alpha_n - \log \alpha_{n+1} = -\beta_0$$

with $\beta_0 \neq 0$. This contradicts Corollary 7.2.1.   □

The corollary implies that numbers like $e.2^{\sqrt{2}}$ and $\pi + \log \alpha$ for any non-zero algebraic α are transcendental. The following is a homogeneous version ($\beta_0 = 0$) of the theorem.

Corollary 7.2.3 *Let* $\alpha_1, \ldots, \alpha_n \in \mathbb{A}\setminus\{0, 1\}$ *and* $\beta_1, \ldots, \beta_n$ *be algebraic numbers with* $1, \beta_1, \ldots, \beta_n$ *linearly independent over the rationals. Then* $\alpha_1^{\beta_1} \cdots \alpha_n^{\beta_n}$ *is transcendental.*

Proof It suffices to show that for any $\alpha_1, \ldots, \alpha_n \in \mathbb{A}\setminus\{0, 1\}$ and $\mathbb{Q}$-linearly independent $\beta_1, \ldots, \beta_n$ that

$$\beta_1 \log \alpha_1 + \cdots + \beta_n \log \alpha_n \neq 0. \tag{7.2}$$

For, then suppose $\alpha_1^{\beta_1} \cdots \alpha_n^{\beta_n}$ were algebraic, say α_{n+1}. Then

$$\beta_1 \log \alpha_1 + \cdots + \beta_n \log \alpha_n - \log \alpha_{n+1} = 0$$

and by hypothesis, $1, \beta_1, \ldots, \beta_n$ are $\mathbb{Q}$-linearly independent which contradicts (7.2) with n replaced by $n + 1$. Thus we need to prove (7.2) which we do by induction on n.

We see that (7.2) is true for $n = 1$. We assume that (7.2) is true for any $m < n$. If $\log \alpha_1, \ldots, \log \alpha_n$ are $\mathbb{Q}$-linearly independent, then the result follows from Theorem 7.1.1. Hence we may assume that $\log \alpha_1, \ldots, \log \alpha_n$ are $\mathbb{Q}$-linearly dependent. Then there are rational numbers $c_1, \ldots, c_n$, not all zero such that

$$c_1 \log \alpha_1 + \cdots + c_n \log \alpha_n = 0.$$

Let us assume without loss of generality that $c_n \neq 0$. Suppose (7.2) does not hold for n. Then together with the above equality, we get

$$(c_n \beta_1 - c_1 \beta_n) \log \alpha_1 + \cdots + (c_n \beta_{n-1} - c_{n-1} \beta_n) \log \alpha_{n-1} = 0.$$

Thus to conclude the proof, it is enough to show that the $(n - 1)$ algebraic numbers $(c_n \beta_1 - c_1 \beta_n), \ldots, (c_n \beta_{n-1} - c_{n-1} \beta_n)$ are linearly independent over $\mathbb{Q}$. Suppose not. Then there exist rational numbers $A_1, \ldots, A_{n-1}$, not all zero such that

$$A_1(c_n \beta_1 - c_1 \beta_n) + \cdots + A_{n-1}(c_n \beta_{n-1} - c_{n-1} \beta_n) = 0$$

which gives

$$c_n A_1 \beta_1 + \cdots + c_n A_{n-1} \beta_{n-1} - (A_1 c_1 + \cdots + A_{n-1} c_{n-1}) \beta_n = 0.$$

Since $\beta_1, \ldots, \beta_n$ are $\mathbb{Q}$-linearly independent, we deduce that $A_1 = \cdots = A_{n-1} = 0$, which is a contradiction. $\qquad\square$

7.3 Applications of the Quantitative Result—Theorem 7.1.2

We shall restrict to the case when $\beta_0 = 0$ and $\beta_i \in \mathbb{Z}$ for $1 \leq i \leq n$.

Corollary 7.3.1 *Let* $\alpha_1, \ldots, \alpha_n \in \mathbb{A} \setminus \{0, 1\}$ *and* $b_1, \ldots, b_n \in \mathbb{Z}$ *such that*

$$\alpha_1^{b_1} \cdots \alpha_n^{b_n} \neq 1.$$

Then

$$|\alpha_1^{b_1} \cdots \alpha_n^{b_n} - 1| \geq (eB)^{-c_{7.3}}$$

where $B = \max(|b_1|, \ldots, |b_n|)$ and $c_{7.3} = c_{7.3}(n, \alpha_1, \ldots, \alpha_n)$.

Proof In the proof, $c_{7.4}$, $c_{7.5}$, $c_{7.6}$ depend on $n, \alpha_1, \ldots, \alpha_n$. For any complex number z, we take $\log z = \log |z| + i \arg z$ with $-\pi < \arg z \leq \pi$. Then

$$\log(1 + z) = \sum_{i=1}^{\infty} \frac{(-1)^{n-1}}{n} z^n \text{ for } z \in \mathbb{C} \text{ with } |z| < 1.$$

Thus when $|z| \leq 1/2$ we get

$$|\log(1 + z)| \leq |z|(1 + |z| + |z|^2 + \cdots) \leq 2|z|.$$

We take $z = \alpha_1^{b_1} \cdots \alpha_n^{b_n} - 1$. If $|z| > 1/2$, the assertion of the corollary follows. So we may assume that $|z| \leq 1/2$. Now

$$\log(1 + z) = b_1 \log \alpha_1 + \cdots + b_n \log \alpha_n + 2k\pi i = b_1 \log \alpha_1 + \cdots + b_n \log \alpha_n + 2k \log(-1)$$

for some $k \in \mathbb{Z}$ as $\log(-1) = \pi i$. We apply Theorem 7.1.2 to get

$$|\log(1 + z)| \geq (e \max(B, |2k|))^{-c_{7.4}}.$$

As $|\log(1 + z)| \leq 2|z| \leq 1$, we have

$$|2k\pi i| \leq 1 + \sum_{i=1}^{n} |\log \alpha_i||b_i| \leq \left(1 + \sum_{i=1}^{n} |\log \alpha_i|\right) B.$$

Hence $|2k| \leq c_{7.5} B$ and we get $|\log(1 + z)| \geq (ec_{7.5}B)^{-c_{7.4}}$. Again using $|\log(1 + z)| \leq 2|z|$ we find

$$|z| \geq (eB)^{-c_{7.6}}.$$

We shall see the implication of the quantitative result for the following well-known conjecture of Pillai. □

Conjecture 7.1 *Let $a, b \in \mathbb{Z}$ and $k \in \mathbb{Z}\backslash\{0\}$ be given. Then there exists a positive number $c_{7.7} = c_{7.7}(a, b, k)$ such that the equation*

$$ax^m - by^n = k \text{ in integers } x > 1, y > 1, m > 1, n > 1 \text{ with } mn \geq 6 \quad (7.3)$$

implies that $\max(x, y, m, n) \leq c_{7.7}$.

This conjecture is still *open* although many special cases are known. We shall consider some cases.

Case 1

Let equation (7.3) hold with $a = b = 1$. Then $\max(m, n) \leq c_{7.8}(x, y, k)$.

Proof Let $B = \max(m, n)$. We may assume without loss of generality that $x^m \geq y^n$. Then by Corollary 7.3.1, we have

$$|1 - x^{-m} y^n| \geq (eB)^{-c_{7.9}} \text{ with } c_{7.9} = c_{7.9}(x, y).$$

Hence

$$|x^m - y^n| \geq \frac{x^m}{(eB)^{c_{7.9}}}.$$

Since $x \geq 2$, $y \geq 2$, we have $x^m \geq 2^B$. Thus

$$|k| = |x^m - y^n| \geq \frac{2^B}{(eB)^{c_{7.9}}}$$

giving

$$2^B \leq c_{7.10} B^{c_{7.9}} \text{ with } c_{7.10} = c_{7.10}(x, y, k).$$

Thus $B \leq c_{7.11}(x, y, k)$.

In particular, equation $3^m - 2^n = 1$ has only finitely many solutions in integers $m > 1, n > 1$. $\qquad\square$

Case 2

Let equation (7.3) hold with $n = m$ and $k > 1$. Then $m \leq c_{7.12}(a, b, k)$.

Proof Note that we may assume that $a \neq 0, b \neq 0$.

First let $a > 0, b < 0$. Let $b = -b'$ with $b' > 0$. Then (7.3) becomes $ax^m + b'y^m = k$ which implies that $ax^m \leq k$ and $b'y^m \leq k$. Thus m, x, y are all bounded by say, k which proves the statement in this case. The case $a < 0, b > 0$ is similar.

Let us now consider $a > 0, b > 0$. We shall assume $x \geq y$. The case $y \geq x$ is similar. Now

$$|k| = |ax^m - by^m| = |ax^m| \left| \frac{b}{a} \left(\frac{y}{x} \right)^m - 1 \right| = ax^m |e^z - 1|$$

where $z = \log(b/a) + m \log(y/x)$. We may assume without loss of generality that $|e^z - 1| < 1/2$. Otherwise, we have $2^m \leq x^m \leq 2|k|/a$ proving the statement.

As a general fact, we claim that there exists an absolute constant $c_{7.13}$ such that

$$|e^z - 1| > c_{7.13}|z| \text{ whenever } |e^z - 1| < 1/2. \tag{7.4}$$

To see this, we note that

$$e^{\Re z} = |e^z| = |e^z - 1 + 1| \le |e^z - 1| + 1 \le 3/2$$

giving

$$\Re z \le \log(3/2) \le 1.$$

Also $|\Im z| \le \pi$. Thus $|z| \le 1 + \pi \le 3\pi/2$. The function $(e^z - 1)/z$ is holomorphic in $|z| \le 3\pi/2$, hence attains its minimum. Thus there exists an absolute constant $c_{7.13}$ satisfying (7.4).

We apply Theorem 7.1.3 to $\Lambda = z$ with $n=2$, $B = m$, $A_1 = h(b/a)=c_{7.14}(a, b) > 0$, $A_2 = h(y/x) = x$. Thus

$$|z| > \exp(-c_{7.15} \log x \log m) \text{ with } c_{7.15} = c_{7.15}(a, b).$$

Thus

$$|k| \ge (ax^m)c_{7.13} \exp(-c_{7.15}(\log x)(\log m))$$

which gives the assertion.

If $a < 0, b < 0$, then writing $a = -a', b = -b'$, the equation becomes $b'y^n - a'x^n = k$ with $a' > 0, b' > 0$. As in the previous case, the assertion follows. $\square$

We proceed to give another application. Let $|m| \ge 2$ be an integer. Denote by $P(m)$ the greatest prime factor of the integer m. We put $P(\pm 1) = 1$.

Theorem 7.3.2 *Let $f(X) \in \mathbb{Z}[X]$ having at least two distinct rational roots. Then*

$$P(f(x)) \to \infty \text{ effectively as } x \to \infty, x \in \mathbb{N}.$$

Remark In fact, the theorem is true when $f(X)$ has two distinct roots. The theorem asserts that for any given $\Delta > 0$, there exists $c_{7.16} = c_{7.16}(\Delta, f) > 0$ such that $P(f(x)) \ge \Delta$ for $x \ge c_{7.16}$. This is what we mean by *effectiveness* here. Suppose $f(X) = X^r, r \ge 1$. Then $P(f(x))$ does not tend to infinity, by allowing x to run through $\{2^n, n = 1, 2, \ldots\}$. Thus the condition that $f(X)$ has two distinct roots is necessary.

Proof Let $f(X) = a_0 X^d + \cdots + a_d, a_i \in \mathbb{Z}$. Then

$$a_0^{d-1} f(X) = (a_0 X)^d + a_1 (a_0 X)^{d-1} + \cdots + a_d a_0^{d-1}.$$

Putting $g(X) = X^d + a_1 X^{d-1} + \cdots + a_d a_0^{d-1}$, we see that $g(X)$ is monic, $g(a_0 X) = a_0^{d-1} f(X)$ and hence $P(f(x)) \to \infty$ if and only if $P(g(x)) \to \infty$. Thus we may assume that $f(X)$ itself is monic.

Let $\Delta > 0$ and x a positive integer satisfying $P(f(x)) \le \Delta$. Then we should show that there exists $c_{7.17} = c_{7.17}(\Delta, f) > 0$ such that $x \le c_{7.17}$. Let the two distinct *rational* roots of $f(X)$ be α_1 and α_2. Since $f(X)$ is monic, these are indeed rational integers. Let

$$c_{7.18} = \max(16, 2|\alpha_1|, 2|\alpha_2|, 4|\alpha_2 - \alpha_1|^2).$$

We assume from now on that $x \geq c_{7.18}$.

Write

$$x - \alpha_1 = p_1^{a_1} \cdots p_s^{a_s}; \quad x - \alpha_2 = q_1^{b_1} \cdots q_t^{b_t}$$

with a_is and b_js positive integers and p_is and q_js primes. By our assumption, $x - \alpha_1 > 0$ and $x - \alpha_2 > 0$. Consider

$$\alpha_2 - \alpha_1 = (x - \alpha_1) - (x - \alpha_2) = p_1^{a_1} \cdots p_s^{a_s} - q_1^{b_1} \cdots q_t^{b_t} \neq 0.$$

Then we get

$$0 \neq |p_1^{a_1} \cdots p_s^{a_s} q_1^{-b_1} \cdots q_t^{-b_t} - 1| = \frac{|\alpha_2 - \alpha_1|}{|x - \alpha_2|} \leq \frac{1}{\sqrt{x}} \tag{7.5}$$

since $x \geq \max(2|\alpha_2|, 4|\alpha_2 - \alpha_1|^2)$. Take

$$z = a_1 \log p_1 + \cdots + a_s \log p_s - b_1 \log q_1 - \cdots - b_t \log q_t.$$

Then by (7.5), we have $z \neq 0$ and

$$0 \neq |e^z - 1| \leq 1/\sqrt{x}. \tag{7.6}$$

Further, z is real. We consider three cases accordingly as $z > 1/2, z < -1/2$ and $|z| \leq 1/2$.

Suppose $z > 1/2$. Then

$$|e^z - 1| > |e^z| - 1 = e^{\Re z} - 1 > e^{1/2} - 1 \geq 1/2.$$

Then by (7.6), we get $x \leq 4$, a contradiction since $x \geq c_{7.18}$.

Suppose $z < -1/2$. Then

$$|e^z - 1| \geq 1 - e^{-1/2}.$$

Along with (7.6), we get a contradiction since $x > 6.5 > 1/(1 - e^{-1/2})^2$.

Suppose $|z| \leq 1/2$. Then it is easy to see that

$$|e^z - 1| \geq |z|/2.$$

Hence, by (7.6), we get

$$|z| < \exp(\log 2 - (\log x)/2) < \exp(-(\log x)/4) \tag{7.7}$$

since $x \geq 16$.

Our final step is to apply Theorem 7.1.4 to get a lower bound for $|z|$ and compare it with the upper bound in (7.7). We need to bound the coefficients a_is, b_js, p_is and q_js in terms of Δ and x.

Now $x - \alpha_1 = p_1^{a_1} \cdots p_s^{a_s}$. Hence for $1 \leq i \leq s$ we have

$$2^{a_i} \leq p_i^{a_i} \leq x - \alpha_1 \leq x^2$$

since $x \geq \max\{2, |\alpha_1|\}$. Thus

$$a_i \leq 2 \log x / \log 2 \leq 3 \log x.$$

Similarly, $b_j \leq 3 \log x$ for $1 \leq j \leq t$.

Since $X - \alpha_1$ divides $f(X)$ and by assumption $P(f(x)) \leq \Delta$, we find that each $p_i \leq \Delta$ and similarly each $q_j \leq \Delta$. Also since s is the number of distinct prime divisors of $x - \alpha_1$, it cannot exceed the largest prime divisor of $f(x)$. Thus $s \leq \Delta$. Similarly, $t \leq \Delta$. Finally, $h(p_i) \leq p_i \leq \Delta$ and $h(q_j) \leq q_j \leq \Delta$. Using these estimates in Theorem 7.1.4 with $n = s + t \leq 2\Delta$ and $d = 1$, we get

$$|z| > \exp(-c_{7.19}(\log \log x))$$

where $c_{7.19} = c_{7.19}(\Delta, f)$. Comparing with the upper bound for $|z|$ in (7.7), we get

$$-c_{7.19}(\log \log x) \leq -(\log x)/4,$$

which gives $x \leq c_{7.20}(\Delta, f)$. Taking $c_{7.17} = \max(c_{7.18}, c_{7.20})$, we get $x \leq c_{7.17}$. $\square$

7.4 Effective Version of Thue's Theorem

In Chap. 5, we saw that equations of the form $f(x, y) = m$ where f is a binary form of degree at least 3 have only finitely many solutions. The method of Thue was *ineffective* in the sense that it was not possible to bound the solutions x and y. Bounding the solutions will also bound the number of solutions. We shall show below how Baker's results can be applied to make the result *effective*, i.e. we will be able to bound the solutions x and y of the Thue equation.

Theorem 7.4.1 *Let* $f(X, Y) \in \mathbb{Z}[X, Y]$ *be an irreducible binary form of degree* $n \geq 3$ *and* m *a non-zero integer. Then there exists an effectively computable number* $c_{7.21} = c_{7.21}(f) > 0$ *such that the equation* $f(x, y) = m$ *with* $x, y \in \mathbb{Z}$ *implies that* $\max(|x|, |y|) \leq (2|m|)^{c_{7.21}}.$

Remark (1) It is well known that Pell's equation $X^2 - 2Y^2 = 1$ has infinitely many solutions. Hence the condition $n \geq 3$ is necessary.

(2) Coefficients of X^n and Y^n are non-zero, otherwise, $f(X, Y)$ is not irreducible.

(3) Suppose $f(X, Y) = a_0 X^n + a_1 X^{n-1} Y + a_2 X^{n-2} Y^2 + \cdots + a_n Y^n$. Then

$$a_0^{n-1} f(X, Y) = (a_0 X)^n + a_1 (a_0 X)^{n-1} Y + a_0 a_2 (a_0 X)^{n-2} Y^2 + \cdots + a_0^{n-1} a_n Y^n = g(a_0 X, Y)$$

where g is a monic polynomial with integral coefficients and $g(a_0 X, Y) = m a_0^{n-1}$. Suppose the theorem is true for g. Then $\max(|a_0 x|, |y|) \leq c_{7.21}$ which implies $\max(|x|, |y|) \leq c_{7.21}$. Hence we may assume that coefficient of X^n in f is 1.

(4) Write $f(X, Y) = Y^n((X/Y)^n + a_1(X/Y)^{n-1} + \cdots + a_n)$. Then $z^n + a_1 z^{n-1} + \cdots + a_n$ is a monic, irreducible polynomial with integral coefficients since f is monic and irreducible. We shall denote this polynomial as $f(z)$.

7.4.1 Proof of Theorem 7.4.1

In the sequel, we denote by $c_{7.22}, c_{7.23}, \ldots$, positive numbers depending only on f. Write
$$f(X, Y) = (X - \alpha^{(1)} Y) \cdots (X - \alpha^{(n)} Y) \text{ with } \alpha^{(1)} = \alpha.$$

Put $\beta = x - \alpha y$. Then β is an algebraic integer in $\mathcal{K} = \mathbb{Q}(\alpha)$, $\beta_i = x - \alpha^{(i)} y$, $1 \leq i \leq n$ and hence $\mathcal{N}(\beta) = |m|$. It is clear that any positive number which depends only on the fundamental units of $\mathcal{K}$ is in fact a number depending only on f. This fact will be used several times without any mention. By Lemma 1.3.8, there exists an associate γ of β such that $\overline{|\gamma|} \leq c_{7.22}$ and

$$\gamma = \zeta \beta \eta_1^{b_1} \cdots \eta_r^{b_r} \ b_i \in \mathbb{Z}, \ \zeta \text{ a root of unity.}$$

Also denominator $d(\gamma) = 1$ since γ is an algebraic integer. Hence $s(\gamma) \leq c_{7.23}$. Let $c' = \max\limits_{1 \leq i, j \leq r} s((\eta_j^{(i)})^{-1})$ and $B = \max\limits_{1 \leq i \leq r} |b_i|$. Then

$$|\beta_i| = \left| \frac{\gamma^{(i)}}{(\eta_1^{(i)})^{b_1} \cdots (\eta_r^{(i)})^{b_r}} \right| \leq c_{7.23}(c')^B \leq c_{7.24}^B. \tag{7.8}$$

From
$$x - \alpha^{(1)} y - \beta_1 = 0; \ x - \alpha^{(2)} y - \beta_2 = 0,$$

we get
$$x = \frac{\alpha^{(1)} \beta_2 - \alpha^{(2)} \beta_1}{\alpha^{(1)} - \alpha^{(2)}}; \quad y = \frac{\beta_2 - \beta_1}{\alpha^{(1)} - \alpha^{(2)}} \tag{7.9}$$

which gives $\max(|x|, |y|) \le c_{7.25} \max(|\beta_1|, |\beta_2|) \le c_{7.26}^B$ by (7.8). Thus if $B \le c_{7.27}$, we see that the theorem is valid.

So we may assume that $B > c_{7.27}$ with $c_{7.27}$ sufficiently large. We have the identity

$$(\alpha^{(2)} - \alpha^{(3)})\beta_1 + (\alpha^{(3)} - \alpha^{(1)})\beta_2 + (\alpha^{(1)} - \alpha^{(2)})\beta_3 = 0.$$

Thus

$$0 \ne \left| \frac{(\alpha^{(2)} - \alpha^{(3)})\beta_1}{(\alpha^{(1)} - \alpha^{(3)})\beta_2} - 1 \right| = \left| \frac{(\alpha^{(1)} - \alpha^{(2)})\beta_3}{(\alpha^{(3)} - \alpha^{(1)})\beta_2} \right|. \tag{7.10}$$

The left-hand side of the above equality is

$$0 \ne \left| \frac{(\alpha^{(2)} - \alpha^{(3)})}{(\alpha^{(1)} - \alpha^{(3)})} \frac{\gamma^{(1)}(\zeta^{(1)})^{-1}(\eta_1^{(1)})^{-b_1} \cdots (\eta_r^{(1)})^{-b_r}}{\gamma^{(2)}(\zeta^{(2)})^{-1}(\eta_1^{(2)})^{-b_1} \cdots (\eta_r^{(2)})^{-b_r}} - 1 \right| = |e^z - 1|$$

where

$$z = \log \frac{\alpha^{(2)} - \alpha^{(3)}}{\alpha^{(1)} - \alpha^{(3)}} + \log \frac{\gamma^{(1)}}{\gamma^{(2)}} + \log \frac{\zeta^{(2)}}{\zeta^{(1)}} + b_1 \log \frac{\eta_1^{(2)}}{\eta_1^{(1)}} + \cdots + b_r \log \frac{\eta_r^{(2)}}{\eta_r^{(1)}} + 2M \log(-1) \ne 0 \tag{7.11}$$

for a suitable integer M. We shall take the principal value of the logarithm so that $\Im z \le \pi$. Since $|\beta_{r_1+i}| = |\beta_{r_1+r_2+i}|$ for $1 \le i \le r_2$ and

$$|\beta_1 \cdots \beta_n| = |m|(\ge 1),$$

we see that $|\beta_i|$ for $1 \le i \le r$ determine $|\beta_i|$ for $r_1 + r_2 \le i \le n$. Thus the above equality implies that there exists I with $1 \le I \le r$ and

$$|\beta_I| \ge 1.$$

We claim that there exists an integer $J, 1 \le J \le n$ and $I \ne J$ such that

$$|\beta_J| \le |m|^{1/(n-1)} e^{-c_{7.28} B}. \tag{7.12}$$

To see this, note that

$$\log \left| \frac{\gamma^{(a)}}{\beta_a} \right| = b_1 \log \left| \eta_1^{(a)} \right| + \cdots + b_r \log \left| \eta_r^{(a)} \right|, \, 1 \le a \le n. \tag{7.13}$$

Let

$$\left| \log \left| \frac{\gamma^{(k)}}{\beta_k} \right| \right| = \max_{1 \le a \le r} \left| \log \left| \frac{\gamma^{(a)}}{\beta_a} \right| \right|.$$

Now (7.13) can be written in the matrix form as

$$T \begin{pmatrix} b_1 \\ \vdots \\ b_r \end{pmatrix} = \begin{pmatrix} \log \left| \frac{\gamma^{(1)}}{\beta_1} \right| \\ \vdots \\ \log \left| \frac{\gamma^{(r)}}{\beta_r} \right| \end{pmatrix}$$

where

$$T = \begin{pmatrix} \log \left| \eta_1^{(1)} \right| \cdots \log \left| \eta_r^{(1)} \right| \\ \vdots \\ \log \left| \eta_1^{(r)} \right| \cdots \log \left| \eta_r^{(r)} \right| \end{pmatrix}.$$

Hence

$$\begin{pmatrix} b_1 \\ \vdots \\ b_r \end{pmatrix} = \frac{\text{Adj } T}{\det T} \begin{pmatrix} \log \left| \frac{\gamma^{(1)}}{\beta_1} \right| \\ \vdots \\ \log \left| \frac{\gamma^{(r)}}{\beta_r} \right| \end{pmatrix}.$$

Thus

$$B \leq c_{7.29} \left\| \log \left| \frac{\gamma^{(k)}}{\beta_k} \right| \right\|$$

giving

$$\left\| \log \left| \frac{\gamma^{(k)}}{\beta_k} \right| \right\| \geq c_{7.30} B.$$

Consider

$$\| \log |\beta_k| \| \geq \left\| \log \left| \frac{\gamma^{(k)}}{\beta_k} \right| \right\| - \log \left| \gamma^{(k)} \right| \geq c_{7.30} B - c_{7.22} \geq c_{7.31} B$$

since $s(\gamma) \leq c_{7.22}$ and B is large. Then

$$\text{either } \log |\beta_k| \leq -c_{7.31} B \text{ or } \log |\beta_k| \geq c_{7.31} B.$$

Suppose $\log |\beta_k| \leq -c_{7.31} B$, then (7.12) holds with $J = k$. So we assume that $\log |\beta_k| \geq c_{7.31} B$. Then

$$|\beta_1 \cdots \beta_{k-1} \beta_{k+1} \cdots \beta_n| = \frac{|m|}{|\beta_k|} \leq |m| e^{-c_{7.31} B}.$$

Let

$$|\beta_i| = \min_{1 \leq j \leq k, j \neq k} |\beta_j|.$$

Then

$$|\beta_i|^{n-1} \le |m|e^{-c_{7.31}B}$$

giving

$$|\beta_i| \le |m|^{1/(n-1)}e^{-c_{7.32}B}$$

showing that (7.12) holds with $J=i$. Note that $I \ne J$ since $|\beta^{(J)}| < |m|^{1/(n-1)}e^{-c_{7.32}B} < 1$ for $B > c_{7.33}\log(2|m|)$.

Now we fix $\beta_2 = \beta_I$ and $\beta_3 = \beta_J$ in (7.10). Then

$$|e^z - 1| = \left| \frac{\alpha^{(1)} - \alpha^{(2)}}{\alpha^{(3)} - \alpha^{(1)}} \frac{\beta_J}{\beta_I} \right| \le c_{7.34}|m|^{1/(n-1)}e^{-c_{7.32}B} < 1/2 \qquad (7.14)$$

for $B > c_{7.35}\log(2|m|)$ with z given by (7.11). As seen in (7.4), we have $|e^z - 1| \ge c_{7.36}|z|$. By Theorem 7.1.3,

$$|z| \ge \exp(-c_{7.37}\log B).$$

Hence

$$|e^z - 1| \ge \exp(-c_{7.38}\log B).$$

Together with the upper bound in (7.14), this implies that $B \le c_{7.39}\log(2|m|)$. Thus by (7.8),

$$|\beta| \le (2|m|)^{c_{7.40}}$$

and hence $|x|$ and $|y|$ are bounded by $(2|m|)^{c_{7.41}}$ by (7.9).   □

As an application of effective Thue's Theorem 7.4.1, we shall show an improvement in Liouville's Theorem 5.2.1 which we recall below.

Let α be an algebraic number of degree $n \ge 2$. Then there exists $c_{5.1} = c_{5.1}(\alpha)$ such that for all integers p and q with $q \ge 1$,

$$\left| \alpha - \frac{p}{q} \right| > \frac{c_{5.1}}{q^n}.$$

We show

Theorem 7.4.2 *Let α be an algebraic number of degree $n \ge 3$. Then there exists $c_{7.42} = c_{7.42}(\alpha) > 0$ and $\kappa = \kappa(\alpha) > 0$, $\kappa < n$ such that for all integers p and q with $q \ge 1$,*

$$\left| \alpha - \frac{p}{q} \right| > \frac{c_{7.42}}{q^\kappa}.$$

Remark The theorems of Thue, Siegel and Roth in Chap. 6 are ineffective since the constant appearing in those theorems corresponding to $c_{7.42}$ is not computable.

Proof of Theorem 7.4.2 We may assume that

$$\left| \alpha - \frac{p}{q} \right| \leq 1.$$

Let $f(X)$ be the minimal polynomial of α. Thus $\deg f = n$. Now

$$\left| f(\alpha) - f\left(\frac{p}{q} \right) \right| = \left| f\left(\frac{p}{q} \right) \right| \geq \frac{1}{q^n}.$$

Let

$$g(p, q) = q^n f(p/q).$$

Then $g(p, q)$ is an irreducible binary form with integral coefficients and $\deg g(p, q) \geq 3$. Hence by Theorem 7.4.1, there exists $c_{7.43} = c_{7.43}(f)$ such that

$$g(p, q) = r \text{ implies that } \max(|p|, |q|) \leq (2|r|)^{c_{7.43}}.$$

Now

$$0 \neq q \leq \max(|p|, |q|) \leq (2|r|)^{c_{7.43}}$$

implies $|r| \geq \dfrac{q^{1/c_{7.43}}}{2}$. Thus

$$\left| f\left(\frac{p}{q} \right) \right| = \frac{|g(p, q)|}{q^n} = \frac{r}{q^n} \geq \frac{1}{2q^{n - \frac{1}{c_{7.43}}}}.$$

Also if $f(X) = a_0 X^n + \cdots + a_n$, then

$$\left| f\left(\frac{p}{q} \right) \right| = \left| f(\alpha) - f\left(\frac{p}{q} \right) \right|$$

$$= \left| a_0 \left(\alpha^n - \left(\frac{p}{q} \right)^n \right) + a_1 \left(\alpha^{n-1} - \left(\frac{p}{q} \right)^{n-1} \right) + \cdots + a_{n-1} \left(\alpha - \frac{p}{q} \right) \right|$$

$$= \left| \alpha - \frac{p}{q} \right| \left| a_0 \left(\frac{\alpha^n - (p/q)^n}{\alpha - (p/q)} \right) + \cdots + a_{n-1} \right|.$$

For any integer t with $2 \leq t \leq n$,

$$\left| \frac{\alpha^t - (p/q)^t}{\alpha - (p/q)} \right| = |\alpha^{t-1} + \alpha^{t-2}(p/q) + \cdots + (p/q)^{t-1}|. \tag{7.15}$$

Since $\left| \alpha - \dfrac{p}{q} \right| \leq 1$, we have $\left| \dfrac{p}{q} \right| \leq 1 + |\alpha|$. Hence from (7.15), we get

$$\left| \frac{\alpha^t - (p/q)^t}{\alpha - (p/q)} \right| \le t(|\alpha| + 1)^{t-1} \le n(|\alpha| + 1)^{n-1}.$$

Therefore,

$$\left| f\left(\frac{p}{q}\right) \right| \le n^2 h(\alpha)(|\alpha| + 1)^{n-1} \left| \alpha - \frac{p}{q} \right|$$

giving

$$\left| \alpha - \frac{p}{q} \right| \ge \left(2q^{n - (1/c_{7.43})} n^2 h(\alpha)(|\alpha| + 1)^{n-1} \right)^{-1} \ge \frac{c_{7.44}}{q^\kappa}$$

with $\kappa = n - \dfrac{1}{c_{7.43}}$.   □

7.5 *p*-Adic Version of Baker's Result and an Application

Analogous results for Baker's theorems in the p-adic metric are due to Kunrui Yu. We state one of his theorems and give an application. Let $\mathbb{Q}_p$ be the completion of $\mathbb{Q}$ under p-adic metric. For any prime p, define the p-adic absolute value $|\ |_p : \mathbb{Z} \to \mathbb{Z}_{\ge 0}$ as follows:

$$|1|_p = 1, |p|_p = 1/p, |q|_p = 1 \text{ for a prime } q \ne p,$$

for every integer factorisation of $\alpha = \pm p_1^{\alpha_1} \cdots p_r^{\alpha_r}$,

$$|\alpha|_p = 1/p^{\alpha_i} \text{ if } p = p_i \text{ for some } 1 \le i \le r \text{ and } |\alpha|_p = 1 \text{ if } p \ne p_i, \text{ for all } 1 \le i \le r.$$

Extend $|\ |_p$ to $\mathbb{Q}$ as follows: If $n = a/b$, then

$$|n|_p = |a|_p/|b|_p.$$

Also $|\ |_p$ defined over $\mathbb{Q}$ satisfies the properties:

(1) $|u|_p \ge 0$ for all $u \in \mathbb{Q}$, $|u|_p = 0$ if and only if $u = 0$.
(2) $|uv|_p = |u|_p |v|_p$.
(3) $|u + v|_p \le \max(|u|_p, |v|_p)$.

Let $\mathcal{K}$ be an algebraic number field. The absolute value $|\ |_p$ can be extended to $\mathcal{K}$ as follows.

Let $\mathfrak{p}$ be a prime ideal of $\mathcal{K}$ which sits over p, i.e. $\mathfrak{p} \cap \mathbb{Z} = p\mathbb{Z}$. Let $\beta \in \mathcal{K}$. Define $|\beta|_p$ as follows. Suppose the principal ideal $[\beta] = \mathfrak{p}^\ell \mathfrak{a}$ for some ideal $\mathfrak{a}$ in $\mathcal{K}$ with $\gcd(\mathfrak{p}, \mathfrak{a}) = 1$ Define $\mathrm{ord}_\mathfrak{p}[\beta] = \ell$. Then

$$|\beta|_\mathfrak{p} = \frac{1}{p^{\mathrm{ord}_\mathfrak{p}[\beta]}}.$$

When $\beta = p$, then $\text{ord}_{\mathfrak{p}}[\beta] = 1$. Hence $|p|_{\mathfrak{p}} = 1/p$. When $\beta = q$, a prime not equal to p then ideal $\mathfrak{p}$ does not occur in the prime factorisation of $[q]$ and we get $\text{ord}_{\mathfrak{p}}[q] = 0$. Therefore $|q|_{\mathfrak{p}} = 1$. Thus $|\ |_{\mathfrak{p}}$ coincides with $|\ |_p$ in the set of integers and hence in $\mathbb{Q}$. We now state a result of Yu.

Theorem 7.5.1 *Let p be a prime number and $\mathcal{K}$ an algebraic number field of degree d. Let $\alpha_1, \ldots, \alpha_n \in \mathcal{K}$ with $s(\alpha_i) \le H$ for $1 \le i \le n$. Further, let $b_1, \ldots, b_n$ be rational integers such that*

$$\alpha_1^{b_1} \cdots \alpha_n^{b_n} \ne 1.$$

Put $B = \max(|b_1|, \ldots, |b_n|)$. Let $\mathfrak{p}$ be a prime ideal of $\mathcal{K}$ which sits over the rational prime p. Then

$$|\alpha_1^{b_1} \cdots \alpha_n^{b_n} - 1|_{\mathfrak{p}} \ge (eB)^{-c_{7.45}}$$

where $c_{7.45} = c_{7.45}(H, n, d, \mathfrak{p})$.

We give an application of the above result.

Theorem 7.5.2 *Let $p_1 < p_2 < \cdots < p_s$ be fixed primes and*

$$S = \{\pm p_1^{a_1} \cdots p_s^{a_s} : a_i \in \mathbb{Z}_{\ge 0}\}$$

i.e., S is the semi-group generated by $p_1, \ldots, p_s$. Let $X_1, X_2, X_3 \in S$ be such that

$$X_1 + X_2 = X_3 \text{ with } \gcd(X_1, X_2) = 1.$$

Then there exists $c_{7.46} = c_{7.46}(p_s, S)$ such that

$$\max(|X_1|, |X_2|) \le c_{7.46}.$$

In other words, the number solutions of $X_1 + X_2 = X_3$ in S is bounded.

Proof Write

$$X_1 = \pm p_1^{a_{11}} \cdots p_s^{a_{1s}}, X_2 = \pm p_1^{a_{21}} \cdots p_s^{a_{2s}}, X_3 = \pm p_1^{a_{31}} \cdots p_s^{a_{3s}}$$

with $a_{ij} \in \mathbb{Z}$, $a_{ij} \ge 0$. Let $Z = \max(|X_1|, |X_2|, |X_3|)$. Observe that for $1 \le i \le s$,

$$2^{a_{1i}} \le p_1^{a_{1i}} \le Z$$

giving $a_{1i} \le 2 \log Z$. Similarly, $a_{2i} \le 2 \log Z$. Consider

$$|X_3|_{p_i} = \frac{1}{p_i^{a_{3i}}} = |X_1 + X_2|_{p_i} = |p_1^{a_{11}} \cdots p_s^{a_{1s}} \pm p_1^{a_{21}} \cdots p_s^{a_{2s}}|_{p_i}. \tag{7.16}$$

Since $\gcd(X_1, X_2) = 1$, p_i cannot divide both X_1 and X_2. Let us assume without loss of generality that $p_i \nmid X_2$. Hence $|X_2|_{p_i} = 1$. Thus from (7.16), we have

$$|X_3|_{p_i} = \frac{|X_3|_{p_i}}{|X_2|_{p_i}} = |\pm p_1^{a_{11}-a_{21}} \cdots p_s^{a_{1s}-a_{2s}} - 1|_{p_i}$$

where $|a_{1i} - a_{2i}| \leq 4 \log Z$. Now applying Theorem 7.5.1 we get

$$|X_3|_{p_i} > \exp(-c_{7.47} \log \log Z)$$

with $c_{7.47} = c_{7.47}(s, p_i)$. On the other hand,

$$|X_3|_{p_i} = \frac{1}{p_i^{a_{3i}}}.$$

Along with the lower bound, this gives

$$p_i^{a_{3i}} < \exp(c_{7.47} \log \log Z).$$

This is true for $1 \leq i \leq s$. Thus

$$|X_3| = \prod_{i=1}^{s} p_i^{a_{3i}} < \exp(c_{7.48} \log \log Z).$$

If $Z = |X_3|$, the above inequality implies that Z is bounded. Suppose $Z = \max(|X_1|, |X_2|)$, say $Z = |X_2|$. Then

$$|X_3| = |X_1 + X_2| = |X_2| \times |\pm p_1^{a_{11}-a_{21}} \cdots p_s^{a_{1s}-a_{2s}} - 1| \geq Z \exp(-c_{7.50} \log \log Z).$$

Again from the upper bound we get

$$\exp(\log Z - c_{7.47} \log \log Z) < \exp(c_{7.48} \log \log Z)$$

implying $Z \leq c_{7.49}$. Hence $\max(|X_1|, |X_2|)$ is bounded. Since $|X_3| \leq |X_1| + |X_2|$, $|X_3|$ is also bounded. Thus the number of solutions of $X_1 + X_2 = X_3$ in S is bounded. $\qquad\square$

abc−Conjecture

In 1983, Masser and Oesterlé formulated *abc* - conjecture as a possible approach to Fermat's last theorem, conjectured by Fermat in 1637, which has been proved by Wiles in 1995 after centuries of tireless efforts and ideas developed by many mathematicians like Euler, Dirichlet, Legendre, Kummer. In twentieth century the problem got an impetus with the works of Frey, Serre, Ribet, Wiles and many others. Before stating *abc* - conjecture, we make some definitions.

Given a sum $a + b = c$ with $a, b, c \in \mathbb{Z}$, coprime and non-zero, define the height h and the radical r of this sum by

$$h = h(a, b, c) = \max(\log|a|, \log|b|, \log|c|); \; r = r(a, b, c) = \sum_{p|abc} \log p$$

where p runs over all prime divisors of a, b and c. For instance, take the following triples (a, b, c)

$$(2, 3, 5) : \; h = \log 5, \; r = \log 30 :$$
$$(9, 16, 25) : \; h = \log 25, \; r = \log 30;$$
$$(3, 125, 128) : \; h = \log 128, \; r = \log 30;$$

$$(19 \times 1307, 7 \times 29^2 \times 31^8, 2^8 \times 3^{22} \times 5^4) : \; h = 36.15..., \; r = 22.36....$$

In the first two examples the height is smaller than the radical. In the next two examples the height is larger than the radical. The *abc* - conjecture says that the height cannot be much larger than the radical.

abc - Conjecture

Let a, b, c be given non-zero, coprime integers such that $a + b = c$ and let $\epsilon > 0$ be given. Then there exists a number $K(\epsilon) > 0$ such that

$$h(a, b, c) \leq r(a, b, c) + \epsilon h(a, b, c) + K(\epsilon).$$

Equivalently, the above inequality can be written as

$$h(a, b, c) \leq \frac{1}{1 - \epsilon} r(a, b, c) + \frac{K(\epsilon)}{1 - \epsilon}.$$

From this we see that if the radical is fixed, i.e. if we consider sums of integers composed of fixed set of prime numbers then there are only finitely many such sums and one can bound the summands. This is Theorem 7.5.2.

abc - Conjecture Implies Fermat's Last Theorem

For a given integer $n \geq 3$, consider $x^n + y^n = z^n$ in positive integers x, y and z. Take $(a, b, c) = (x^n, y^n, z^n)$. Then $h = n \log z$ and

$$r = \sum_{p|xyz} \log p \leq \log xyz < 3 \log z.$$

Applying *abc* - conjecture with $\epsilon = 1/2$, we get

$$n \log z \leq 6 \log z + 2K(1/2).$$

Since we know the equation has no solution for $n = 3, 4, 5$ and 6, we may assume that $n > 6$. Thus

$$(n - 6) \log z \leq 2K(1/2)$$

which implies that n, z and hence x and y are bounded and these finitely many values are left for direct verification.

We give another application of abc - conjecture. In 1909, Wieferich proved *if p is a prime satisfying*

$$2^{p-1} \not\equiv 1 \pmod{p^2}, \tag{7.17}$$

then the equation $x^p + y^p = z^p$ has no non-trivial integral solutions satisfying $p \nmid xyz$.

It is still *not known* if there are infinitely many primes p satisfying (7.17). We shall show this under abc - conjecture. This was proved by Silverman [9] in 1988. We need the following notion.

We say that an integer N is *square full* if a prime q divides N, then q^2 divides N. For example, $N = 2^8 \cdot 3^{17}$ is a square-full number. In fact, any positive integer N can be written as

$$N = u_N v_N$$

where u_N is a square-free part of N, and v_N is a square-full part of N so that $(u_N, v_N) = 1$.

Theorem 7.5.3 *Under abc - conjecture, there are infinitely many primes p satisfying* (7.17).

Proof For any positive integer n, write $2^n - 1 = u_n v_n$ where u_n is the square-free part of $2^n - 1$ and v_n is the square-full part of $2^n - 1$. Thus $(u_n, v_n) = 1$. We claim the following. □

Claim *Assume that abc - conjecture is true. Then as $n \to \infty$, the factor u_n is unbounded.*

Suppose u_n is bounded. Consider the equality

$$(2^n - 1) + 1 = 2^n.$$

Taking $a = 2^n - 1, b = 1, c = 2^n$, we get $h(a, b, c) = n \log 2$ and

$$r(a, b, c) = \log(2 \prod_{p|(2^n-1)} p) \leq \log 2 + \log u_n + (\log v_n)/2.$$

Then by the abc - conjecture, for any $\epsilon > 0$, there exists $K(\epsilon)$ such that

$$n \log 2 \leq \frac{\log 2 + \log u_n + (\log v_n)/2}{1 - \epsilon} + \frac{K(\epsilon)}{1 - \epsilon}.$$

On the other hand,

$$n \log 2 > \log(2^n - 1) = \log u_n + \log v_n.$$

Comparing with the upper bound, we see that

$$(1/2 - \epsilon) \log v_n \leq \epsilon \log u_n + \log 2 + K(\epsilon).$$

Fixing $\epsilon = 1/4$, we get that v_n is bounded since u_n is bounded. This implies $2^n - 1$ is bounded which is not true by taking n arbitrarily large. This proves the claim.

By the claim, as u_n is square free, we conclude that there are infinitely many primes p such that p divides u_n for some positive integer n. Hence, in order to prove the theorem, it is enough to show that every prime divisor p of u_n for $n \geq 1$ satisfies (7.17). Let $p|u_n$ for some $n \geq 1$. Then $p^2 \nmid 2^n - 1$.

Let d be the order of 2 (mod p). Thus $2^d \equiv 1 \pmod{p}$ and $2^h \not\equiv 1 \pmod{p}$ for any $h < d, d|n$ and $d|(p-1)$. So let $n = de$ and $p - 1 = df$. Suppose $2^d \equiv 1 \pmod{p^2}$. Then $2^d = 1 + kp$ with $p|k$. Hence

$$2^n = 2^{de} = (1 + pk)^e \equiv 1 + pke \equiv 1 \pmod{p^2},$$

a contradiction. Thus $2^d \not\equiv 1 \pmod{p^2}$. Then $2^d = 1 + mp$ for some integer m with $p \nmid m$. Hence,

$$2^{p-1} = 2^{df} = (1 + mp)^f \equiv 1 + fmp \pmod{p^2}.$$

Since $f|(p-1)$ and $p \nmid m$ we see that $2^{p-1} \not\equiv 1 \pmod{p^2}$ which proves the theorem. $\qquad\square$

Exercise

(1) Show that $\exp(\alpha\pi + \beta)$ is transcendental if $\alpha, \beta \in \mathbb{A}, \beta \neq 0$.
(2) Show that the series

$$1 - \frac{1}{4} + \frac{1}{7} - \frac{1}{10} + \cdots$$

is transcendental.
(3) Is π^e transcendental? What can be said about $\log \log 2, \log^2 2 + \log^2 3$?
(4) Show that Fermat equation $x^n + y^n = z^n$ has no solution for $n = 3, 4, 5$ and 6.
(5) Assuming *abc* - conjecture, show that there are only finitely many consecutive cube full numbers.
(6) Assuming *abc* - conjecture, show that the number of Wieferich primes not exceeding X is $\gg \log X / \log \log X$.

Notes

There are innumerable articles on the application of Baker's results, both qualitative and quantitative, in different areas of number theory. In the past decade, a theorem of Baker, Birch and Wirsing [10], in which Baker's theory of linear forms was used, found many applications. Starting with a paper of Adhikari et al. [11] and of Murty and Saradha [12], their theorem was applied to prove the transcendence of many infinite series. We refer to the book of Murty and Rath [13] and an article of Saradha and Sharma [14] for various results in this connection, including transcendence of some integrals.

The equation

$$x^m - y^n = 1 \tag{7.18}$$

is known as *Catalan equation*. In 1844, Catalan conjectured that the only two consecutive perfect powers are 2^3 and 3^2, i.e. the only positive integral solutions of (7.18) is $(x, y, m, n) = (3, 2, 2, 3)$. Using linear forms in logarithms, in 1976, Tijdeman [15] showed that *the equation* (7.18) *in positive integers* x, y, m, n *implies that* $\max(x, y, m, n)$ *is bounded by an effectively computable absolute constant*. In 2002, Catalan conjecture was solved by Mihǎilescu [16] using cyclotomic fields and Galois modules.

In 1897, Störmer used Pell's equation to show that $P(x(x + 1)) \to \infty$ as $x \to \infty$. He showed that there are only 23 pairs $(x, x + 1)$ which are composed of fixed primes and he gave *explicitly* all the 23 pairs. Thue, in 1908, noted that, his method would give finiteness of the number of such pairs $(x, x + 1)$. But his method was *ineffective* in the sense that one cannot furnish the actual pairs as Störmer did. On the other hand, we saw in Theorem 7.3.2 that Baker's method gives effective result even if $x(x + 1)$ is replaced by a polynomial $P(x)$ having at least two distinct roots. But, computing all the x values needs more analysis and techniques.

In 1964 Baker [17] used properties of hyper-geometric series to obtain effective results for approximation of certain fractional powers of rationals by rationals. For instance, it was shown that

$$\left| 2^{1/3} - \frac{p}{q} \right| > \frac{10^{-6}}{q^{2.955}}.$$

Since then using the Padé approximants occurring in the hyper-geometric method has become an important tool in Diophantine approximation. Further, approximation of more than one algebraic number simultaneously has also been studied.

References

1. A. Baker, Linear forms in the logarithms of algebraic numbers. Mathematika **13**, 204–216 (1966); II **14**, 102–107 (1967); III **14**, 220–228 (1967)
2. A. Baker, *Transcendental Number Theory*. Cambridge Tracts (1975)
3. A. Baker, G. Wüstholz, *Logarithmic Forms and Diophantine Geometry*. Cambridge Tracts (2007)
4. M. Laurent, M. Mignotte, Y. Nesterenko, Formes linéaires en deux logarithmes et déterminants d'interpolation. J. Number Theory **55**, 285–321 (1995)
5. C.D. Bennett, J. Blass, A.M.W. Glass, D.B. Meronk, R.P. Steiner, Linear forms in the logarithms of three positive rational integers. J. Theo. Nombr. Bordeaux **9**, 97–136 (1997)
6. E.M. Matveev, An explicit lower bound for a homogeneous rational linear form in logarithms of algebraic numbers. Izv. Math. **62**, 81–136 (1998)
7. Y. Bugeaud, *Linear Forms in Logarithms and Applications*. IRMA Lectures in Mathematics and Theoretical Physics, vol. 28 (2018)

8. T.N. Shorey, R. Tijdeman, *Exponential Diophantine Equations*. Cambridge Tracts (1986 and re-printed in 2008)
9. J. Silverman, Wieferich's criterion and the ABC- conjecture. J. Number Theory **30**, 226–237 (1988)
10. A. Baker, B.J. Birch, E.A. Wirsing, On a problem of Chowla. J. Number Theory **5**, 224–236 (1973)
11. S.D. Adhikari, N. Saradha, T.N. Shorey, R. Tijdeman, Transcendental infinite sums. Indag Math. (N. S.) **12**(1), 1–14 (2001)
12. M.R. Murty, N. Saradha, Transcendental values of the digamma function. J. Number Theory **125**, 298–318 (2007)
13. M. Ram Murty, P. Rath, *Transcendental Numbers* (Springer, Berlin, 2014), 217 pp
14. N. Saradha, D. Sharma, *Arithmetic Nature of Some Infinite Series and Integrals*. Contemporary Mathematics **655**, 191–207 (2015)
15. R. Tijdeman, On the equation of Catalan. Acta Arith. **29**, 197–209 (1976)
16. P. Mihăilescu, Catalan's conjecture: another old Diophantine problem solved. Bull. Am. Math. Soc. **41**, 43–57 (2013)
17. A. Baker, Rational approximations to $2^{1/3}$ and other algebraic numbers. Quart J. Math. Oxford **15**, 375–383 (1964)

Chapter 8
Baker's Theorem

The city you're dreaming of it's at the end of this road

–Lal Ded

We begin with some basic tools necessary for the proof of Theorem 7.1.1 in Sect. 8.1. First, Theorem 7.1.1 is reduced to an equivalent statement; see Theorem 8.1.2. In Sect. 8.1.1, we derive a simple, but useful, non-trivial lower bound for a non-vanishing linear form in logarithms of algebraic numbers with bounded coefficients. Section 8.1.2 provides construction of an augmentative polynomial. In Sect. 8.1.3, we give the construction of the auxiliary polynomial $\Phi(Z_0, \ldots, Z_{n-1})$ in several variables which generalises the function of a single complex variable employed by Gelfond. Basic estimates on Φ are shown in Sect. 8.1.4. The main difficulty is in the interpolation techniques. Usually the order of the derivatives is increased while leaving the points of interpolation fixed. Baker used a special extrapolation procedure in which the range of interpolation points is extended while the order of the derivatives is reduced, and the absolute values of these derivatives are shown to be very small. See Sects. 8.1.5 and 8.1.6.

In Sect. 8.2, all the tools of Sect. 8.1 are combined in an ingenious way. The auxiliary polynomial $\Phi(Z_0, \ldots, Z_{n-1})$ at $Z_0 = \cdots = Z_{n-1} = z$ is expressed as an exponential polynomial

$$\Phi(z, \ldots, z) = \sum p_\alpha z^{\nu_\alpha} e^{\psi_\alpha z} \not\equiv 0 \text{ with } p_\alpha \in \mathbb{Z}.$$

Then using the augmentative polynomial, the coefficients p_α of $\Phi(z, \ldots, z)$ are expressed in terms of the derivatives of Φ at $z = 0$. Using the smallness of the derivatives, it is shown that

$$|p_\alpha| < 1$$

© Springer Nature Singapore Pte Ltd. 2020
S. Natarajan and R. Thangadurai, *Pillars of Transcendental Number Theory*,
https://doi.org/10.1007/978-981-15-4155-1_8

for every α which implies that $\Phi(z) \equiv 0$, a contradiction to the construction.

We refer to the original papers of Baker [1] and his book [2] for the presentation here. The reader may also refer to [3] for some fascinating exposition.

8.1   Ground Work for the Proof of Baker's Theorem

A Reduction

Let us recall Baker's result from Theorem 7.1.1.

Theorem 8.1.1 *Let $\alpha_1, \ldots, \alpha_n \in \mathbb{A} \setminus \{0, 1\}$, $\beta_0 \in \mathbb{A}$ and $\beta_1, \ldots, \beta_n \in \mathbb{A} \setminus \{0\}$. Assume that*

$$\log \alpha_1, \ldots, \log \alpha_n \text{ are linearly independent over } \mathbb{Q}.$$

Then $\Lambda := \beta_0 + \beta_1 \log \alpha_1 + \cdots + \beta_n \log \alpha_n \neq 0$. In other words,

$$1, \log \alpha_1, \ldots, \log \alpha_n$$

are linearly independent over $\mathbb{A}$.

In order to prove Theorem 8.1.1, we first show that it is enough to prove the following statement.

Theorem 8.1.2 *Let $\alpha_1, \ldots, \alpha_n$ be algebraic numbers such that $\log \alpha_1, \ldots, \log \alpha_n$ are $\mathbb{Q}$-linearly independent. Suppose $\beta_0, \beta_1, \ldots, \beta_n$ are non-zero algebraic numbers. Then there exist non-negative integers $\lambda_1, \ldots, \lambda_n$ such that*

$$e^{\lambda_n \beta_0} \alpha_1^{\lambda_1 + \lambda_n \beta_1} \ldots \alpha_{n-1}^{\lambda_{n-1} + \lambda_n \beta_{n-1}} \neq \alpha_1^{\lambda_1} \cdots \alpha_n^{\lambda_n}. \tag{8.1}$$

We shall show that Theorem 8.1.2 implies Theorem 8.1.1. Assume that (8.1) holds. We need to prove that for any $\beta_0, \beta_1, \ldots, \beta_n \in \mathbb{A}$, not all zero, we have

$$\beta_0 + \beta_1 \log \alpha_1 + \cdots + \beta_n \log \alpha_n \neq 0.$$

Since β_js are not all zero, by rearranging the indices, if necessary, we can assume that $\beta_n \neq 0$. Since $\beta_n \in \mathbb{A} \setminus \{0\}$, we have $\beta_n^{-1} \in \mathbb{A} \setminus \{0\}$. By letting, $\gamma_i = -\beta_i/\beta_n$ for $0 \leq i < n$, we need to prove that for any $\gamma_0, \gamma_1, \ldots, \gamma_{n-1} \in \mathbb{A}$, not all zero, we have

$$\gamma_0 + \gamma_1 \log \alpha_1 + \cdots + \gamma_{n-1} \log \alpha_{n-1} \neq \log \alpha_n. \tag{8.2}$$

Suppose (8.2) is not true. Then there exist $\gamma_0, \ldots, \gamma_{n-1} \in \mathbb{A}$, not all zero, for which

$$\gamma_0 + \gamma_1 \log \alpha_1 + \cdots + \gamma_{n-1} \log \alpha_{n-1} = \log \alpha_n.$$

By exponentiating both sides, we get,

$$e^{\gamma_0} \alpha_1^{\gamma_1} \cdots \alpha_{n-1}^{\gamma_{n-1}} = \alpha_n.$$

Therefore, for all non-negative integer λ_n, we have

$$e^{\lambda_n \gamma_0} \alpha_1^{\lambda_n \gamma_1} \cdots \alpha_{n-1}^{\lambda_n \gamma_{n-1}} = \alpha_n^{\lambda_n}.$$

Multiplying both sides by $\alpha_1^{\lambda_1} \cdots \alpha_{n-1}^{\lambda_{n-1}}$, we get, for all non-negative integers $\lambda_1, \ldots, \lambda_n$ that

$$e^{\lambda_n \gamma_0} \alpha_1^{\lambda_1 + \lambda_n \gamma_1} \cdots \alpha_{n-1}^{\lambda_{n-1} + \lambda_n \gamma_{n-1}} = \alpha_1^{\lambda_1} \cdots \alpha_n^{\lambda_n},$$

which is a contradiction to (8.1). $\square$

Thus we need to prove Theorem 8.1.2. We assume that the assertion of the theorem is not true and arrive at a contradiction. This means we can assume that for all non-negative integers $\lambda_0, \lambda_1, \ldots, \lambda_n$, we have

$$e^{\lambda_n \beta_0} \alpha_1^{\lambda_1 + \lambda_n \beta_1} \cdots \alpha_{n-1}^{\lambda_{n-1} + \lambda_n \beta_{n-1}} = \alpha_1^{\lambda_1} \cdots \alpha_n^{\lambda_n}. \tag{8.3}$$

Therefore, it follows, by raising both sides to ℓth power that for all positive integers ℓ, we have

$$e^{\lambda_n \beta_0 \ell} \alpha_1^{\ell(\lambda_1 + \lambda_n \beta_1)} \cdots \alpha_{n-1}^{\ell(\lambda_{n-1} + \lambda_n \beta_{n-1})} = \alpha_1^{\ell \lambda_1} \cdots \alpha_n^{\ell \lambda_n}. \tag{8.4}$$

We shall denote by $c_{8,\ldots} = c_{8,\ldots}(\alpha_1, \ldots, \alpha_n, \beta_0, \ldots, \beta_{n-1})$ effectively computable positive numbers depending only on $\alpha_1, \ldots, \alpha_n, \beta_0, \ldots, \beta_{n-1}$. By this dependence, we mean any quantity which can be determined once $\alpha_1, \ldots, \alpha_n, \beta_0, \ldots, \beta_{n-1}$ are given, for instance, when n, $\deg(\alpha_i)$, $d(\alpha_i)$, $h(\alpha_i)$ for $1 \leq i \leq n$ and $\deg(\beta_j)$, $d(\beta_j)$, $h(\beta_j)$ for $0 \leq j < n$ are given. Let $\mathcal{K} = \mathbb{Q}(\alpha_1, \ldots, \alpha_n, \beta_0, \ldots, \beta_{n-1})$ and $[\mathcal{K} : \mathbb{Q}] = \nu$. We put $a_i = d(\alpha_i)$, $e_i = d(1/\alpha_i)$ for $1 \leq i \leq n$ and $b_j = d(\beta_j)$ for $0 \leq j \leq n - 1$.

8.1.1 A Lower Bound for a Non-vanishing Linear Form

Lemma 8.1.3 *Let* $\alpha_1, \ldots, \alpha_n$ *be non-zero algebraic numbers such that* $\log \alpha_1, \ldots, \log \alpha_n$ *are* $\mathbb{Q}$-*linearly independent complex numbers. Let* $T > 0$ *be any real number. Suppose* $t_1, \ldots, t_n$ *are, not all zero, integers with* $|t_i| \leq T$, *for* $1 \leq i \leq n$. *Then, there exists* $c_{8.1}$ *such that*

$$|t_1 \log \alpha_1 + t_2 \log \alpha_2 + \cdots + t_n \log \alpha_n| \geq c_{8.1}^{-T}.$$

Proof Since $\log \alpha_1, \ldots, \log \alpha_n$ are $\mathbb{Q}$-linearly independent, we see that

$$t_1 \log \alpha_1 + \cdots + t_n \log \alpha_n \neq 0.$$

We want to prove a lower bound for this number. So, let

$$\Omega = t_1 \log \alpha_1 + \cdots + t_n \log \alpha_n \in \mathbb{C}^*.$$

Then

$$e^{\Omega} = e^{t_1 \log \alpha_1 + \cdots + t_n \log \alpha_n} = \alpha_1^{t_1} \cdots \alpha_n^{t_n} \in \mathbb{A}.$$

Without loss of generality, we may assume that

$$t_1, \ldots, t_k \geq 0 \text{ and } t_{k+1}, \ldots, t_n < 0.$$

Let $t_j = -s_j$ with $s_j > 0$ for all $j = k+1, k+2, \ldots, n$. Consider the complex number

$$\omega = a_1^{t_1} \cdots a_k^{t_k} e_{k+1}^{s_{k+1}} \cdots e_n^{s_n} \left(\alpha_1^{t_1} \cdots \alpha_k^{t_k} \alpha_{k+1}^{-s_{k+1}} \cdots \alpha_n^{-s_n} - 1 \right) \in \mathbb{A}.$$

Then, ω is an algebraic integer in $\mathcal{K}$. Hence

$$\deg(\omega) \leq \nu$$

and

$$|\sigma(\omega)| \leq c_{8.2}^{T} \tag{8.5}$$

for any conjugate $\sigma(\omega)$ of ω.

Suppose $\omega = 0$. Then $e^{\Omega} = 1$ giving $\Omega = 2\pi i m$ for some integer $m \neq 0$, as $\Omega \neq 0$. Therefore, we get

$$|\Omega| = |2\pi m| \geq 2\pi > 1,$$

which proves the assertion of the lemma by taking $c_{8.1} = 1$.

Now consider $\omega \neq 0$. Then

$$\prod_{\sigma} |\sigma(\omega)| = |\mathcal{N}(\omega)| \geq 1$$

giving by (8.5) that

$$|\omega| \geq \frac{1}{\prod_{\sigma(\omega) \neq \omega} |\sigma(\omega)|} \geq (c_{8.2}^{\nu-1})^{-T}.$$

Thus we get

$$|e^{\Omega} - 1| = \frac{|\omega|}{a_1^{t_1} \cdots a_k^{t_k} e_{k+1}^{s_{k+1}} \cdots e_n^{s_n}} \geq \frac{1}{(c_{8.2}^{\nu-1} c_{8.3})^{T}}.$$

Since for any complex number z, we know that $|e^z - 1| \leq |z| e^{|z|}$ and $|\Omega| \leq c_{8.4} T$, we get

$$|\Omega| = \frac{|\Omega| e^{|\Omega|}}{e^{|\Omega|}} \geq \frac{|e^{\Omega} - 1|}{e^{|\Omega|}} \geq \frac{1}{c_{8.5}^{T}}$$

where $c_{8.5} = c_{8.2}^{\nu-1} c_{8.3} e^{c_{8.4}}$. $\square$

8.1.2   A Special Augmentative Polynomial

Lemma 8.1.4 *Let R and S be given positive integers. Let $\sigma_0, \sigma_1, \ldots, \sigma_{R-1}$ be given distinct complex numbers. Define*

$$\sigma = \max\{1, |\sigma_n| \ : \ 0 \le n \le R - 1\} \text{ and } \rho = \min\{1, |\sigma_i - \sigma_j| \ : \ 0 \le i < j \le R - 1\}.$$

Let $(r, s) \in [0, R - 1] \times [0, S - 1]$ be a given pair of integers. Then there exists a polynomial

$$W(z) = \sum_{j=0}^{RS-1} c_j z^j \in \mathbb{C}[z]$$

of degree at most $RS - 1$ such that

(a) $|c_j| \le \left(\dfrac{2\sigma}{\rho}\right)^{RS}$ *for $0 \le j < RS$;*

(b) $\dfrac{d^i}{dz^i}(W(z))|_{z=\sigma_j} = \begin{cases} 0 & \text{if } (i, j) \ne (s, r) \\ 1 & \text{if } (i, j) = (s, r). \end{cases}$

Proof Fix a pair (r, s). Consider

$$W(z) = a_0(z - \sigma_r)^s \prod_{\substack{m=0 \\ m \ne r}}^{R-1} (z - \sigma_m)^S$$

where

$$a_0 = \left(s! \prod_{\substack{m=0 \\ m \ne r}}^{R-1} (\sigma_r - \sigma_m)^S\right)^{-1}.$$

Clearly, $W(z)$ is a polynomial with complex coefficients of degree at most $RS - 1$. Using Leibnitz's formula, it can be seen that

$$\frac{d^i}{dz^i}(W(z))|_{z=\sigma_j} = 0 \text{ for } i < S \text{ and } i \ne r$$

and

$$\frac{d^s}{dz^s}(W(z))|_{z=\sigma_r} = a_0 s! \prod_{\substack{m=0 \\ m \ne r}}^{R-1} (\sigma_r - \sigma_m)^S = 1,$$

by the definition of a_0. Hence (b) is satisfied. Now, we estimate the coefficients of $W(z)$. Write $W(z) = c_0 + c_1 z + \cdots + c_{RS-1} z^{RS-1}$. Note that

$$W(z) \ll |a_0|(z + \sigma)^{RS}.$$

Hence

$$|c_m| \le |a_0| \times \text{Coefficient of } z^m \text{ in } (z + \sigma)^{RS}, 0 \le m < RS.$$

Note that the coefficient of z^m in $(z + \sigma)^{RS}$ is equal to

$$\binom{RS}{m} \sigma^m < (\sigma + 1)^{RS}$$

and

$$|a_0| = \left(s! \prod_{\substack{m=0 \\ m \ne r}} |\sigma_r - \sigma_m|^s \right)^{-1} < \rho^{-S(R-1)},$$

by the definition of ρ. Thus we get

$$|c_m| < \frac{(\sigma + 1)^{RS}}{\rho^{S(R-1)}} = \left(\frac{\sigma + 1}{\rho} \right)^{RS} \rho^S \le \left(\frac{2\sigma}{\rho} \right)^{RS},$$

as $\rho \le 1$.   □

8.1.3   Construction of the Auxiliary Function

The next lemma describes the auxiliary function Φ which is fundamental for the proof of Theorem 8.1.2. For any function $F(Z_0, \ldots, Z_{n-1})$, by $F_{m_0,\ldots,m_{n-1}}(Z_0, \ldots, Z_{n-1})$ we mean

$$\left(\frac{\partial}{\partial Z_0} \right)^{m_0} \cdots \left(\frac{\partial}{\partial Z_{n-1}} \right)^{m_{n-1}} F(Z_0, \ldots, Z_{n-1}).$$

When we evaluate the above expression at $Z_0 = z_0, \ldots, Z_{n-1} = z_{n-1}$ we simply write it as $F_{m_0,\ldots,m_{n-1}}(z_0, \ldots, z_{n-1})$. Further, if $z_0 = \cdots = z_{n-1} = z$ we write as $F_{m_0,\ldots,m_{n-1}}(z)$.

Lemma 8.1.5 *Let the equation (8.4) hold. Then there exists a constant $c_{8.6}$ with the following property. For any integer $h \ge c_{8.6}$ and $L = \left[h^{2-\frac{1}{4n}} \right]$, there exist integers $p(\lambda_0, \ldots, \lambda_n)$ for $0 \le \lambda_0, \ldots, \lambda_n \le L$, not all zero, with*

$$|p(\lambda_0, \ldots, \lambda_n)| < e^{h^3}$$

such that the function

$$\Phi(z_0, z_1, \ldots, z_{n-1}) = \sum_{\lambda_0=0}^{L} \cdots \sum_{\lambda_n=0}^{L} p(\lambda_0, \ldots, \lambda_n) z_0^{\lambda_0} e^{\lambda_n \beta_0 z_0} \alpha_1^{(\lambda_1+\lambda_n\beta_1)z_1} \cdots \alpha_{n-1}^{(\lambda_{n-1}+\lambda_n\beta_n)z_{n-1}}$$

satisfies

$$\Phi_{m_0, m_1, \ldots, m_{n-1}}(\ell) = 0$$

for all $1 \le \ell \le h$ and for all non-negative integers $m_0, m_1, \ldots, m_{n-1}$ with $m_0 + m_1 + \cdots + m_{n-1} \le h^2$.

Proof Let h be a large positive integer, and we prove the existence of such a constant $c_{8.6}$ satisfying the assertion.

System of Equations
Consider

$$\Phi_{m_0, m_1, \ldots, m_{n-1}}(\ell) = 0$$

for $1 \le \ell \le h$ and for non-negative integers $m_0, m_1, \ldots, m_{n-1}$ with $m_0 + m_1 + \cdots + m_{n-1} \le h^2$. This is a system of equations in $(L+1)^{n+1}$ unknowns $p(\lambda_0, \ldots, \lambda_n)$, and the number of equations is at most $h(h^2+1)^n$. Note that

$$(L+1)^{n+1} \ge \left(h^{2-\frac{1}{4n}}\right)^{n+1} = h^{2n+2-\frac{n+1}{4n}} > \nu(\nu+1)h(h^2+1)^n \qquad (8.6)$$

by taking $h > (\nu(\nu+1)2^n)^{4n/(3n-1)}$, say. Here $\nu = [\mathcal{K} : \mathbb{Q}]$.

Simplification of a Typical Equation
Consider

$$\Phi_{m_0, m_1, \ldots, m_{n-1}}(\ell) = 0.$$

By the definition of $\Phi_{m_0, \ldots, m_{n-1}}(z_0, \ldots, z_{n-1})$, we get,

$$\sum_{\lambda_0=0}^{L} \cdots \sum_{\lambda_n=0}^{L} p(\lambda_0, \ldots, \lambda_n) \frac{\partial^{m_0}}{\partial z_0^{m_0}} \left(z_0^{\lambda_0} e^{\lambda_n \beta_0 z_0}\right)|_{z_0=\ell} \frac{\partial^{m_1}}{\partial z_1^{m_1}} \left(\alpha_1^{(\lambda_1+\lambda_n\beta_1)z_1}\right)|_{z_1=\ell} \cdots$$

$$\cdots \frac{\partial^{m_{n-1}}}{\partial z_{n-1}^{m_{n-1}}} \left(\alpha_{n-1}^{(\lambda_{n-1}+\lambda_n\beta_{n-1})z_{n-1}}\right)|_{z_{n-1}=\ell} = 0. \qquad (8.7)$$

By Leibnitz's formula, we find

$$\frac{\partial^{m_i}}{\partial z_i^{m_i}} \left(\alpha_i^{(\lambda_i+\lambda_n\beta_i)z_i}\right)|_{z_i=\ell} = (\lambda_i + \lambda_n\beta_i)^{m_i} (\log \alpha_i)^{m_i} \alpha_i^{(\lambda_i+\lambda_n\beta_i)\ell} \text{ for } 1 \le i \le n-1$$

and

$$\frac{\partial^{m_0}}{\partial z_0^{m_0}} \left(z_0^{\lambda_0} e^{\lambda_n\beta_0 z_0}\right)|_{z_0=\ell} = \sum_{\mu_0=0}^{m_0} \binom{m_0}{\mu_0} \frac{\partial^{\mu_0}}{\partial z_0^{\mu_0}} (z_0^{\lambda_0})|_{z_0=\ell} \frac{\partial^{m_0-\mu_0}}{\partial z_0^{m_0-\mu_0}} \left(e^{\lambda_n\beta_0 z_0}\right)|_{z_0=\ell}$$

$$= \sum_{\mu_0=0}^{m_0} \binom{m_0}{\mu_0} \lambda_0 (\lambda_0 - 1) \cdots (\lambda_0 - \mu_0 + 1) \ell^{\lambda_0 - \mu_0} (\lambda_n \beta_0)^{m_0 - \mu_0} e^{\lambda_n \beta_0 \ell}.$$

Therefore, Eq. (8.7) becomes,

$$\sum_{\lambda_0=0}^{L} \cdots \sum_{\lambda_n=0}^{L} p(\lambda_0, \ldots, \lambda_n) \left(\sum_{\mu_0=0}^{m_0} \binom{m_0}{\mu_0} \left(\prod_{i=0}^{\mu_0-1} (\lambda_0 - i) \right) \ell^{\lambda_0-\mu_0} (\lambda_n \beta_0)^{m_0-\mu_0} e^{\lambda_n \beta_0 \ell} \right)$$

$$\tag{8.8}$$

$$\times \prod_{i=1}^{n-1} (\log \alpha_i)^{m_i} (\lambda_i + \lambda_n \beta_i)^{m_i} \alpha_i^{(\lambda_i + \lambda_n \beta_i)\ell} = 0.$$

Since $\log \alpha_1, \ldots, \log \alpha_{n-1}$ are $\mathbb{Q}$- linearly independent, we see that

$$(\log \alpha_1)^{m_1} \cdots (\log \alpha_{n-1})^{m_{n-1}} \neq 0$$

and independent of the above sum. Thus Eq. (8.8) reduces to

$$\sum_{\substack{\lambda_0,\ldots,\lambda_n \\ =0}}^{L} p(\lambda_0, \ldots, \lambda_n) q(m_0, \lambda_0, \lambda_n, \ell, \beta_0) \left(\prod_{i=1}^{n-1} (\lambda_i + \lambda_n \beta_i)^{m_i} \right) \left(e^{\lambda_n \beta_0 \ell} \prod_{i=1}^{n-1} \alpha_i^{(\lambda_i + \lambda_n \beta_i)\ell} \right) = 0$$

$$\tag{8.9}$$

where

$$q(m_0, \lambda_0, \lambda_n, \ell, \beta_0) = \sum_{\mu_0=0}^{m_0} \binom{m_0}{\mu_0} \lambda_0 (\lambda_0 - 1) \cdots (\lambda_0 - \mu_0 + 1) \ell^{\lambda_0-\mu_0} (\lambda_n \beta_0)^{m_0-\mu_0}.$$

Application of (8.4)
Using Eq. (8.4) in (8.9) we get

$$\sum_{\substack{\lambda_0,\ldots,\lambda_n \\ =0}}^{L} p(\lambda_0, \ldots, \lambda_n) q(m_0, \lambda_0, \lambda_n, \ell, \beta_0) \left(\prod_{i=1}^{n-1} (\lambda_i + \lambda_n \beta_i)^{m_i} \right) \alpha_1^{\lambda_1 \ell} \cdots \alpha_n^{\lambda_n \ell} = 0.$$

$$\tag{8.10}$$

Recall that $a_i = d(\alpha_i)$ for $1 \leq i \leq n$ and $b_j = d(\beta_j)$ for $0 \leq j \leq n - 1$. We multiply Eq. (8.10) by $(a_1 a_2 \cdots a_n)^{L\ell} b_0^{m_0} \cdots b_{n-1}^{m_{n-1}}$ to get

$$\sum_{\substack{\lambda_0,\ldots,\lambda_n \\ =0}}^{L} p(\lambda_0, \ldots, \lambda_n) q'(m_0, \lambda_0, \lambda_n, \ell, \beta_0) \prod_{i=1}^{n-1} (b_i \lambda_i + \lambda_n (b_i \beta_i))^{m_i} \prod_{i=1}^{n} a_i^{(L-\lambda_i)\ell} \prod_{i=1}^{n} (a_i \alpha_i)^{\lambda_i \ell} = 0,$$

$$\tag{8.11}$$

where

$$q'(m_0, \lambda_0, \lambda_n, \ell, \beta_0) = \sum_{\mu_0=0}^{m_0} \binom{m_0}{\mu_0} \lambda_0 (\lambda_0 - 1) \cdots (\lambda_0 - \mu_0 + 1) \ell^{\lambda_0-\mu_0} b_0^{\mu_0} (\lambda_n b_0 \beta_0)^{m_0-\mu_0}.$$

Note that the coefficients of $p(\lambda_0, \cdots, \lambda_n)$ in (8.11) are algebraic integers in $\mathcal{K}$.

Estimation of the Coefficients

(a) $\displaystyle\prod_{i=1}^{n} |a_i \alpha_i|^{\lambda_i \ell} \le c_{8.7}^{Lh}$

(b) $\displaystyle\prod_{i=1}^{n-1} |b_i \lambda_i + \lambda_n b_i \beta_i|^{m_i} \le L^{(n-1)h^2} c_{8.8}^{(n-1)h^2} \le L^{c_{8.9}h^2}.$

(c) $\displaystyle\prod_{i=1}^{n-1} a_i^{(L-\lambda_i)\ell} \le c_{8.10}^{Lh}.$

(d)

$$|q'(m_0, \lambda_0, \lambda_n, \ell, \beta_0)| = \left| \sum_{\mu_0=0}^{m_0} \binom{m_0}{\mu_0} \lambda_0 (\lambda_0 - 1) \cdots (\lambda_0 - \mu_0 + 1) \ell^{\lambda_0 - \mu_0} b_0^{\mu_0} (\lambda_n b_0 \beta_0)^{m_0 - \mu_0} \right|$$

$$\le (m_0 + 1) 2^{m_0} \lambda_0^{\mu_0} \ell^{\lambda_0} \lambda_n^{m_0} c_{8.11}^{m_0}$$

$$\le L^{c_{8.12}h^2} h^L.$$

Combining (a)–(d) the coefficients in Eq. (8.11) are bounded by

$$e^{c_{8.13}Lh + c_{8.14}h^2 \log L}.$$

In fact, the above bound is true for any conjugate of the coefficients. Therefore, by Lemma 3.1.3(ii) with $v = (L+1)^{n+1}$, $w = h(h^2+1)^n$, and (8.6), we get

$$|p(\lambda_0, \ldots, \lambda_n)| \le c_{8.15}(L+1)^{n+1} e^{c_{8.13}Lh + c_{8.14}h^2 \log L}.$$

Final Estimate

We have

(i) $(n+1)\log(L+1) \le (n+1)\log 2 + (n+1)\left(2 - \dfrac{1}{4n}\right)\log h,$

(ii) $Lh \le h^{3-\frac{1}{4n}},$

(iii) $h^2 \log L \le \left(2 - \dfrac{1}{4n}\right) h^2 \log h.$

Hence we get

$$|p(\lambda_0, \ldots, \lambda_n)| \le e^{c_{8.16}h^{3-1/4n}} < e^{h^3}$$

whenever $h > c_{8.16}^{4n}$. Thus the lemma is proved by taking $h > c_{8.6}$ where

$$c_{8.6} = \max(c_{8.16}^{4n}, (\nu(\nu+1)2^n)^{4n/(3n-1)}).$$

□

8.1.4 Basic Estimates Relating to Φ

Lemma 8.1.6 Let $h \geq c_{8.6}$ and Φ as constructed in Lemma 8.1.5. Let $m_0, m_1, \ldots,$ m_{n-1} be non-negative integers such that $m_0 + m_1 + \cdots + m_{n-1} \leq h^2$. Let

$$f_{m_0,\ldots,m_{n-1}}(z) := \frac{\partial^{m_0}}{\partial z_0^{m_0}} \cdots \frac{\partial^{m_{n-1}}}{\partial z_{n-1}^{m_{n-1}}} (\Phi(z_0, \ldots, z_{n-1})) \, |_{z_0 = \cdots z_{n-1} = z}.$$

Then there exist $c_{8.17}$ and $c_{8.18}$ satisfying the following properties:

(a) For any $z \in \mathbb{C}$,

$$\left| f_{m_0,\ldots,m_{n-1}}(z) \right| \leq c_{8.17}^{h^3 + L|z|};$$

(b) For any integer ℓ, we have either $f_{m_0,\ldots,m_{n-1}}(\ell) = 0$ or

$$\left| f_{m_0,\ldots,m_{n-1}}(\ell) \right| > c_{8.18}^{-h^3 - L|\ell|}.$$

Proof Proof of (a) From (8.8), by replacing ℓ by z we see that

$$f_{m_0,\ldots,m_{n-1}}(z) = \sum_{\substack{\lambda_0,\ldots,\lambda_n \\ =0}}^{L} p(\lambda_0, \ldots, \lambda_n) q(m_0, \lambda_0, \lambda_n, z, \beta_0) \left(\prod_{i=1}^{n-1} (\lambda_i + \lambda_n \beta_i)^{m_i} \right)$$

$$\left(\prod_{i=1}^{n-1} (\log \alpha_i)^{m_i} \right) \left(\alpha_1^{\lambda_1} \cdots \alpha_n^{\lambda_n} \right)^z, \tag{8.12}$$

where

$$q(m_0, \lambda_0, \lambda_n, z, \beta_0) = \sum_{\mu_0=0}^{m_0} \binom{m_0}{\mu_0} (\lambda_0(\lambda_0 - 1) \cdots (\lambda_0 - \mu_0 + 1)) z^{\lambda_0 - \mu_0} (\lambda_n \beta_0)^{m_0 - \mu_0}.$$

Now, we shall estimate the individual terms to get an upper bound.

(a') By Lemma 8.1.5, we know that

$$|p(\lambda_0, \ldots, \lambda_n)| \leq e^{h^3}.$$

(b') Consider

$$|q(m_0, \lambda_0, \lambda_n, z, \beta_0)| \leq \sum_{\mu_0=0}^{m_0} \binom{m_0}{\mu_0} |\lambda_0||\lambda_0 - 1| \cdots |\lambda_0 - \mu_0 + 1||z|^{\lambda_0 - \mu_0} |\lambda_n \beta_0|^{m_0 - \mu_0}$$

$$\leq (c_{8.19})^{m_0} L^{m_0} (1 + |z|)^{\lambda_0}$$

$$\leq c_{8.19}^{h^2} L^{h^2} (1 + |z|)^{L}.$$

(c')

$$\left| \prod_{i=1}^{n-1} (\lambda_i + \lambda_n \beta_i)^{m_i} \right| \leq L^{m_1 + \cdots + m_{n-1}} c_{8.20}^{m_1 + \cdots + m_{n-1}}$$

$$\leq L^{h^2} c_{8.20}^{h^2}.$$

(d')

$$\left| \prod_{i=1}^{n-1} (\log \alpha_i)^{m_i} \right| \leq \max_i |\log \alpha_i|^{m_1 + \cdots + m_{n-1}}$$

$$\leq c_{8.21}^{h^2}.$$

(e')

$$\left| \left(\alpha_1^{\lambda_1} \cdots \alpha_n^{\lambda_n} \right)^z \right| \leq \left| \alpha_1^{\lambda_1} \cdots \alpha_n^{\lambda_n} \right|^{|z|}$$

$$\leq c_{8.22}^{L|z|}.$$

Thus, by $(a') - -(e')$, we get

$$|f_{m_0,\ldots,m_{n-1}}(z)| \leq e^{h^3} L^{c_{8.23} h^2 + L} c_{8.24}^{L|z|}$$

$$\leq c_{8.25}^{h^3 + L|z|},$$

which proves (a). $\qquad \qquad \square$

Proof of (b)

Let ℓ be a given integer. We assume that $f_{m_0,\ldots,m_{n-1}}(\ell) \neq 0$. Let

$$g_0(\ell) = \frac{f_{m_0,\ldots,m_{n-1}}(\ell)}{(\log \alpha_1)^{m_1} \cdots (\log \alpha_{n-1})^{m_{n-1}}}.$$

and

$$g(\ell) = b_0^{m_0} \cdots b_{n-1}^{m_{n-1}} a_1^{L\ell} \cdots a_n^{L\ell} g_0(\ell).$$

Then $g(\ell)$ is a non-zero algebraic integer in $\mathcal{K}$. Hence

$$\mathcal{N}(g(\ell)) = g(\ell) \prod_{\sigma \neq 1} \sigma(g(\ell)) \geq 1,$$

where the product is over all the embeddings σ of $\mathcal{K}$ into $\mathbb{C}$ except the identity map. Note that for any embedding $\sigma : \mathcal{K} \to \mathbb{C}$, we have

$$\sigma(g(\ell)) = \sum_{\substack{\lambda_0,\ldots,\lambda_n \\ =0}}^{L} p(\lambda_0,\ldots,\lambda_n) q(m_0,\lambda_0,\ell,\sigma(\beta_0)) \left(\prod_{i=1}^{n-1} (b_i\lambda_i + \lambda_n b_i \sigma(\beta_i))^{m_i} \right) \times$$

$$\times \prod_{i=1}^{n} (a_i\sigma(\alpha_i))^{\lambda_i\ell} a_i^{(L-\lambda_i)\ell}.$$

By (8.12), we see that

$$|\sigma(g(\ell))| \leq \frac{\left| \prod_{i=1}^{n} a_i^{L\ell} \prod_{j=0}^{n-1} b_j^{m_j} \right|}{\prod_{i=1}^{n-1} |\log\alpha_i|^{m_i}} |f_{m_0,\ldots,m_{n-1}}(\ell)|.$$

Therefore, using the upper bound for $|f_{m_0,\ldots,m_{n-1}}(\ell)|$ from (a), we get

$$|\sigma(g(\ell))| \leq c_{8.26}^{h^3+L|\ell|}.$$

Thus

$$|g(\ell)| \geq \frac{1}{\prod_{\sigma\neq 1} |\sigma(g(\ell))|}$$

$$\geq \left(\prod_{\sigma\neq 1} c_{8.26}^{h^3+L|\ell|} \right)^{-1}$$

$$= c_{8.26}^{(\nu-1)(-h^3-L|\ell|)}$$

$$= c_{8.27}^{-h^3-L|\ell|}.$$

Therefore, we get

$$|f_{m_0,\ldots,m_{n-1}}(\ell)| \geq c_{8.28}^{-h^3-L|\ell|},$$

as desired. $\square$

8.1.5 Extrapolation Technique to Get More Zeros

By Lemma 8.1.5, we know that

$$f_{m_0,\ldots,m_{n-1}}(\ell) = 0$$

whenever $1 \leq \ell \leq h$ and $m_0 + \cdots + m_{n-1} \leq h^2$. Now we will increase the range of ℓ at the cost of a reduction in $m_0 + \cdots + m_{n-1}$.

Lemma 8.1.7 *There exists $c_{8.29}$ such that for any integer $h > c_{8.29}$ the following holds. Let J be an integer with $0 \le J \le (8n)^2$. For all integers ℓ with $1 \le \ell \le h^{1+\frac{J}{8n}}$ and for all non-negative integers $m_0, \ldots, m_{n-1}$ with $m_0 + \cdots + m_{n-1} \le h^2/2^J$, we have*

$$f_{m_0,\ldots,m_{n-1}}(\ell) = 0.$$

Proof We assume $h > c_{8.6}$ where $c_{8.6}$ is the number occurring in Lemma 8.1.5 so that the results of Lemmas 8.1.5 and 8.1.6 hold.

We prove the lemma by induction on J. When $J = 0$, we have $h^{1+\frac{J}{8n}} = h$ and $h^2/2^J = h^2$. Therefore the conclusion follows from Lemma 8.1.5. Hence, we shall assume that the result holds when $J = k$ and prove the result for $J = k + 1$.

We take $m_0, \ldots, m_{n-1}$ to be non-negative integers such that

$$m_0 + \cdots + m_{n-1} \le h^2/2^{k+1}. \tag{8.13}$$

Then $m_0 + \cdots + m_{n-1} \le h^2/2^k$. Let ℓ be any integer with $1 \le \ell \le h^{1+\frac{k}{8n}}$. By induction hypothesis, it follows that

$$f_{m_0,\ldots,m_{n-1}}(\ell) = 0.$$

Hence, we may assume that

$$h^{1+\frac{k}{8n}} < \ell \le h^{1+\frac{k+1}{8n}}. \tag{8.14}$$

Suppose that

$$f_{m_0,\ldots,m_{n-1}}(\ell) \ne 0$$

with $m_0, \ldots, m_{n-1}$ and ℓ satisfying (8.13) and (8.14). We compute lower and upper bounds for this non-zero quantity which contradict each other. This will prove the lemma.

Lower Bound for $|f_{m_0,\ldots,m_{n-1}}|$

By Lemma 8.1.6,

$$\left| f_{m_0,\ldots,m_{n-1}}(\ell) \right| \ge c_{8.18}^{-h^3 - L|\ell|}$$
$$\ge c_{8.30}^{-h^{3+k/(8n)}} \tag{8.15}$$

where $c_{8.30}$ can be taken as $c_{8.18}^2$. The upper bound for $|f_{m_0,\ldots,m_{n-1}}(\ell)|$ comes after many steps.

Construction of an Entire Function with Many Zeros

Let

$$R_k = \left[h^{1+\frac{k}{8n}} \right]; \quad S_k = \left[\frac{h^2}{2^k} \right].$$

By (8.14) and (8.13), we see that ℓ satisfies $R_k < \ell \le R_{k+1}$ and $m_0, \ldots, m_{n-1}$ satisfies $m_0 + \cdots + m_{n-1} \le S_{k+1}$. Take any non-negative integer $m \le S_{k+1}$. Writing $m = j_0 + \cdots + j_{n-1}$ with each j_t non-negative integer, we see that

$$
\begin{aligned}
(m_0 + j_0) + \cdots + (m_{n-1} + j_{n-1}) &= (m_0 + \cdots + m_{n-1}) + (j_0 + \cdots + j_{n-1}) \\
&\le S_{k+1} + m \le 2S_{k+1} \\
&\le 2\left[\frac{h^2}{2^{k+1}}\right] \le \frac{h^2}{2^k}.
\end{aligned}
$$

Therefore, by induction hypothesis, for any integer r with $1 \le r \le R_k$, we have

$$
f_{m_0+j_0,\ldots,m_{n-1}+j_{n-1}}(r) = 0 \tag{8.16}
$$

with $j_0 + \cdots + j_{n-1} = m \le S_{k+1}$. Let

$$
g(z) = f_{m_0,\ldots,m_{n-1}}(z)
$$

and put $\frac{d^m}{dz^m}(g(z)) = g_m(z)$. Since

$$
g(r) = f_{m_0,\ldots,m_{n-1}}(r) = \frac{\partial^{m_0}}{\partial z_0^{m_0}} \cdots \frac{\partial^{m_{n-1}}}{\partial z_{n-1}^{m_{n-1}}} \Phi(z_0, \ldots, z_{n-1})|_{z_0=\cdots=z_{n-1}=r}
$$

by Leibnitz's formula, we get,

$$
\begin{aligned}
g_m(r) &= \sum_{\substack{j_0,\ldots,j_{n-1} \\ j_0+\cdots+j_{n-1}=m}} \binom{m}{j_0,\ldots,j_{n-1}} \frac{\partial^{j_0}}{\partial z_0^{j_0}} \cdots \frac{\partial^{j_{n-1}}}{\partial z_{n-1}^{j_{n-1}}} \left(\Phi_{m_0,\ldots,m_{n-1}}(z_0,\ldots,z_{n-1})\right)\Big|_{z_0=\cdots=z_{n-1}=r} \\
&= \sum_{\substack{j_0,\ldots,j_{n-1} \\ j_0+\cdots+j_{n-1}=m}} \binom{m}{j_0,\ldots,j_{n-1}} \Phi_{m_0+j_0,\ldots,m_{n-1}+j_{n-1}}(z_0,\ldots,z_{n-1})\Big|_{z_0=\cdots=z_{n-1}=r} \\
&= \sum_{\substack{j_0,\ldots,j_{n-1} \\ j_0+\cdots+j_{n-1}=m}} \binom{m}{j_0,\ldots,j_{n-1}} f_{m_0+j_0,\ldots,m_{n-1}+j_{n-1}}(r). \tag{8.17}
\end{aligned}
$$

Thus, by Eqs. (8.16) and (8.17), we see that the entire function $g(z)$ has zeros at $z = 1, 2, \ldots, R_k$ of order $\ge S_{k+1}$.

Application of Maximum–Minimum Modulus Principle
Let

$$
F(z) = [(z-1)(z-2)\cdots(z-R_k)]^{S_{k+1}}.
$$

Then, $g(z)/F(z)$ is an entire function. Let

$$
R = R_{k+1}h^{\frac{1}{8n}},
$$

Thus $R > 2R_k$ by taking $h > 2^{8n}$, say. Let D_R be the closed disc $|z| \le R$. By the maximum–minimum modulus principle, we have

$$\left| \frac{g(z)}{F(z)} \right| \le \frac{\max_{|z|=R} |g(z)|}{\min_{|z|=R} |F(z)|} \text{ for } z \in D_R.$$

Thus, for any $z \in D_R$, we get

$$|g(z)| \le \frac{|F(z)|(\max_{|z|=R} |g(z)|)}{\min_{|z|=R} |F(z)|}.$$

Upper Bound for $\max_{|z|=R} |g(z)|$

By Lemma 8.1.6 (a),

$$\max_{|z|=R} |g(z)| \le c_{8.17}^{h^3+LR}. \tag{8.18}$$

Lower Bound for $\min_{|z|=R} |F(z)|$

Since $\ell > R_k$, $F(\ell) \ne 0$. Further

$$|F(\ell)| = (|\ell - 1||\ell - 2| \cdots |\ell - R_k|)^{S_{k+1}} \le R_{k+1}^{R_k S_{k+1}} \tag{8.19}$$

since $|\ell - j| \le \ell \le R_{k+1}$. As $R > 2R_k$, we get for $1 \le j \le R_k$, $|z - j| > \frac{R}{2}$. Therefore,

$$\min_{|z|=R} |F(z)| \ge \left(\frac{R}{2} \right)^{R_k S_{k+1}}. \tag{8.20}$$

From (8.18)–(8.20), we get

$$|g(\ell)| \le c_{8.17}^{h^3+LR} (2R_{k+1}/R)^{R_k S_{k+1}}.$$

Note that

$$R_k S_{k+1} \le \frac{h^{3+\frac{k}{8n}}}{2^{k+1}} \tag{8.21}$$

and

$$c_{8.17}^{h^3+LR} \le c_{8.17}^{h^3+h^{3+\frac{k}{8n}}} \le c_{8.31}^{h^{3+\frac{k}{8n}}} \tag{8.22}$$

by taking $c_{8.31} = c_{8.17}^2$, say. Using (8.21) and (8.22), we get

$$|g(\ell)| \le c_{8.31}^{h^{3+\frac{k}{8n}}} \left(\frac{2}{h^{1/8n}} \right)^{(h^{3+\frac{k}{8n}})/(2^{k+1})} \tag{8.23}$$

$$\le c_{8.32}^{h^{3+\frac{k}{8n}}} h^{-(h^{3+\frac{k}{8n}})/(2^{k+4}n)}$$

where $c_{8.32}$ can be taken as $2c_{8.31}$.

Final Contradiction
Comparing (8.15) and (8.23) we have

$$\left(h^{3+\frac{k}{8n}}\right)\log c_{8.32} - \frac{h^{3+\frac{k}{8n}}}{n2^{k+4}}\log h \geq -\left(h^{3+\frac{k}{8n}}\right)\log c_{8.30}$$

i.e.

$$\left(h^{3+\frac{k}{8n}}\right)(\log c_{8.32} + \log c_{8.30}) \geq \left(h^{3+\frac{k}{8n}}\right)\frac{\log h}{n2^{k+4}}$$

or

$$\log h \leq n2^{k+4}\log(c_{8.33})$$

with $c_{8.33} = c_{8.30}c_{8.32}$. Taking $h > \max\left(c_{8.33}^{n2^{(8n)^2+4}}, 2^{8n}, c_{8.6}\right)$, we get the final contradiction. $\qquad\square$

8.1.6 Smallness of Derivatives

Let $f(z) = \Phi(z, \ldots, z)$ with Φ as in Lemma 8.1.5.

Lemma 8.1.8 *There exists a constant $c_{8.34}$ such that for all integers $h > c_{8.34}$ the following property holds. For all integers j with $0 \leq j \leq h^{8n}$, we have*

$$\left|\frac{d^j}{dz^j}(f(z))|_{z=0}\right| < e^{-h^{8n}}.$$

Proof Let $h > c_{8.29}$ so that Lemma 8.1.7 is valid. Let

$$X = h^{8n} \text{ and } Y = \left[\frac{h^2}{2^{(8n)^2}}\right].$$

Then by Lemma 8.1.7 we have

$$f_{m_0,\ldots,m_{n-1}}(\ell) = 0 \text{ for } 1 \leq \ell \leq X$$

and for all non-negative integers $m_0, \ldots, m_{n-1}$ with $m_0 + \cdots + m_{n-1} \leq Y$. That is, f has zeros at $z = 1, 2, \ldots, X$ of order at least Y.

Application of Maximum–Minimum Modulus Principle
Let $R = Xh^{1/(8n)}$. Then $R > 2X$ since $h > c_{8.29} \geq 2^{8n}$. Let C be the circle $|z| = R$ and D_R the closed disc $|z| \leq R$. Put

$$F(z) = \frac{f(z)}{[(z-1)(z-2)\cdots(z-X)]^Y}.$$

Then $F(z)$ is an entire function. Therefore, by maximum–minimum modulus principle, for all $z \in \bar{D}_R$, we get

$$|F(z)| \le \frac{\max_{|z|=R}|f(z)|}{\min_{|z|=R}(|z-1|\cdots|z-X|)^Y}.$$

This implies that

$$|f(z)| \le \frac{\max_{|z|=R}|f(z)|}{\min_{|z|=R}(|z-1|\cdots|z-X|)^Y}\,(|z-1|\cdots|z-X|)^Y.$$

First we estimate $\max_{|z|=R}|f(z)|$. By Lemma 8.1.6 (a), we get

$$\max_{|z|=R}|f(z)| \le c_{8.17}^{h^3+LR} \tag{8.24}$$

$$\le c_{8.17}^{2h^{2+8n}}$$

$$\le c_{8.17}^{2^{(8n)^2+1}XY} \le c_{8.35}^{XY}.$$

Next since $X < R/2$, we get

$$\min_{|z|=R}(|z-1|\cdots|z-X|)^Y > (R-X)^{XY} \tag{8.25}$$

$$> \left(\frac{R}{2}\right)^{XY}.$$

Lastly, for all $|z| \le X$, we have

$$(|z-1|\cdots|z-X|)^Y \le (2X)^{XY}. \tag{8.26}$$

Combining (8.24)–(8.26), we obtain for all $|z| \le X$,

$$|f(z)| \le (4c_{8.35}X/R)^{XY} < \exp(-XY) \tag{8.27}$$

for $h > (4ec_{8.35})^{8n}$.

Application of Cauchy Integral Formula

For any integer $0 \le j \le h^{8n}$ and circle $C' : |z| = X$, we have by Cauchy Integral formula and (8.27) that

$$|f^{(j)}(0)| \le \frac{j!}{2\pi} \int_{C'} \frac{|f(z)|}{|z|^{j+1}} dz$$
$$\le j^j \frac{e^{-XY}}{X^j}$$
$$\le e^{-XY}.$$

Observe that $XY = h^{8n+2}/2^{(8n)^2} > h^{8n}$ if $h > 2^{(8n)^2/2}$. This completes the proof of the lemma by taking $c_{8.34} = \max(c_{8.29}, (4ec_{8.35})^{8n}, 2^{(8n)^2/2})$. □

8.2 Proof of Baker's Theorem

We assume that the hypotheses of Theorem 8.1.2 hold. Then Lemmas 8.1.5–8.1.8 are valid. The main strategy now is to show that the inequalities in Lemma 8.1.8 cannot all be valid simultaneously.

Unique Representation of Integers
Let $n \ge 1$ and $L \ge 1$ be integers. Put $S = L + 1$ and $R = S^n$. It is well known that any non-negative integer α can be represented uniquely in base S as

$$\alpha = \lambda_0 + \lambda_1 S + \cdots + \lambda_\ell S^\ell \tag{8.28}$$

with $0 \le \lambda_i \le L$ for $0 \le i \le \ell$ for some integer $\ell \ge 0$. It is easy to see that every integer α with $0 \le \alpha \le S^{n+1} - 1$ can be represented uniquely as in (8.28) with $\ell \le n$ and we write

$$\alpha = (\lambda_0, \lambda_1, \ldots, \lambda_n).$$

Set

$$\nu_\alpha = \lambda_0, \quad p_\alpha = p(\lambda_0, \ldots, \lambda_n), \quad \psi_\alpha = \lambda_1 \log \alpha_1 + \cdots + \lambda_n \log \alpha_n. \tag{8.29}$$

Cardinality of the set $\{\psi_\alpha\}$ with $0 \le \alpha \le RS - 1$.

Lemma 8.2.1 *Let*
$$\mathcal{S} = \{\psi_\alpha : 0 \le \alpha \le RS - 1\}.$$

Then
$$|\mathcal{S}| = S^n.$$

Proof Let $(\lambda_0, \ldots, \lambda_n)$ and $(\kappa_0, \ldots, \kappa_n) \in [0, L]^{n+1}$ be two tuples. Then $(\lambda_0, \ldots, \lambda_n) \ne (\kappa_0, \ldots, \kappa_n)$ if and only if

either $\lambda_0 \ne \kappa_0$ or $(\lambda_1, \ldots, \lambda_n) \ne (\kappa_1, \ldots, \kappa_n)$.

Therefore, the number of distinct elements $(\lambda_0, \ldots, \lambda_n) \in [0, L]^{n+1}$ such that $\lambda_0 = \kappa_0$ is precisely $(L+1)^n = R$. Consider $\alpha = (\kappa_0, \lambda_1, \ldots, \lambda_n)$ and $\beta = (\kappa_0, \kappa_1, \ldots, \kappa_n)$ with $\alpha \neq \beta$. Then we claim that

$$\psi_\alpha \neq \psi_\beta.$$

Suppose not. Then

$$(\lambda_1 - \kappa_1) \log \alpha_1 + \cdots + (\lambda_n - \kappa_n) \log \alpha_n = 0.$$

Since for some i with $1 \leq i \leq n$, we have $\lambda_i - \kappa_i \neq 0$, the above equality implies that $\log \alpha_1, \ldots, \log \alpha_n$ are $\mathbb{Q}$-linearly dependent, a contradiction which proves the claim. Thus αs with fixed λ_0 correspond to distinct ψ_α. Thus $|S| = R$. $\square$

Connecting to Augmentative Polynomial

We list the distinct elements of S as $\psi_0, \psi_1, \ldots, \psi_{R-1}$. Let r and s be two given integers with $0 \leq r \leq R - 1$ and $0 \leq s \leq S - 1$. Then by Lemma 8.1.4 there exists a polynomial

$$W(z) = \sum_{i=0}^{RS-1} c_i z^i \in \mathbb{C}[z] \tag{8.30}$$

of degree at most $RS - 1$ and satisfying

(a) $|c_i| < (2\sigma/\rho)^{RS}$;

(b) $\dfrac{d^j}{dz^j}(W(z))|_{z=\psi_i} = \begin{cases} 1 & \text{if } (i, j) = (r, s) \\ 0 & \text{if } (i, j) \neq (r, s) \end{cases}$

where $\sigma = \max\{1, |\psi_i| : 0 \leq i \leq R - 1\}$ and $\rho = \min\{1, |\psi_i - \psi_j| : 0 \leq i < j \leq R - 1\}$.

Lemma 8.2.2 *Let ψ_α, ν_α and p_α for $\alpha \in [0, RS - 1]$ be as defined in (8.29) and $W(z)$ be the polynomial given in (8.30). Then*

$$\left(\frac{d}{dz}\right)^{\nu_\alpha}(W(z))|_{z=\psi_\alpha} = \sum_{j=0}^{RS-1} c_j \left(\frac{d}{dz}\right)^j (z^{\nu_\alpha} e^{\psi_\alpha z})|_{z=0}.$$

Proof Consider

$$\left(\frac{d}{dz}\right)^{\nu_\alpha}(W(z))|_{z=\psi_\alpha} = \left(\frac{d}{dz}\right)^{\nu_\alpha}\left(\sum_{k=0}^{RS-1} c_k z^k\right)|_{z=\psi_\alpha}$$

$$= \sum_{k=\nu_\alpha}^{RS-1} c_k(k(k-1)\ldots(k-\nu_\alpha+1))\psi_\alpha^{k-\nu_\alpha} \tag{8.31}$$

Now, for any positive integer k, by Leibnitz's formula, we have,

$$\left(\frac{d}{dz}\right)^k \left(z^{\nu_\alpha} e^{\psi_\alpha z}\right) = \sum_{j=0}^{k} \binom{k}{j} \left(\frac{d}{dz}\right)^j (z^{\nu_\alpha}) \left(\frac{d}{dz}\right)^{k-j} (e^{\psi_\alpha z})$$

$$= \sum_{j=0}^{k} \binom{k}{j} (\nu_\alpha(\nu_\alpha - 1)\ldots(\nu_\alpha - j + 1)) z^{\nu_\alpha - j} \psi_\alpha^{k-j} e^{\psi_\alpha z}$$

Therefore,

$$\left(\frac{d}{dz}\right)^k \left(z^{\nu_\alpha} e^{\psi_\alpha z}\right)|_{z=0} = \text{constant term in the above expression} \qquad (8.32)$$

$$= \binom{k}{\nu_\alpha} (\nu_\alpha!) \psi_\alpha^{k-\nu_\alpha}$$

$$= (k(k-1)\ldots(k - \nu_\alpha + 1) \psi_\alpha^{k-\nu_\alpha}.$$

The lemma follows from (8.31) and (8.32).   □

Connecting to the Auxiliary Polynomial Φ

Lemma 8.2.3 Let ψ_α, ν_α and p_α for $\alpha \in [0, RS - 1]$ be as defined in (8.29). Let $\Phi(z_0, \ldots, z_{n-1})$ be the polynomial defined in Lemma 8.1.5. Then we have

$$\Phi(z, z, \ldots, z) = \sum_{\alpha=0}^{RS-1} p_\alpha z^{\nu_\alpha} e^{\psi_\alpha z}.$$

Proof By Lemma 8.1.5, we have

$$\Phi(z_0, \ldots, z_{n-1}) = \sum_{\substack{\lambda_0,\ldots,\lambda_n \\ =0}}^{L} p(\lambda_0, \ldots, \lambda_n) z_0^{\lambda_0} e^{\lambda_n \beta_0 z_0} \alpha_1^{(\lambda_1+\lambda_n\beta_1)z_1} \ldots \alpha_{n-1}^{(\lambda_{n-1}+\lambda_n\beta_{n-1})z_{n-1}}$$

and hence

$$\Phi(z, \ldots, z) = \sum_{\substack{\lambda_0,\ldots,\lambda_n \\ =0}}^{L} p(\lambda_0, \ldots, \lambda_n) z^{\lambda_0} \left(e^{\lambda_n \beta_0} \alpha_1^{\lambda_1+\lambda_n\beta_1} \ldots \alpha_{n-1}^{\lambda_{n-1}+\lambda_n\beta_{n-1}}\right)^z.$$

Therefore, by Eq. (8.3), we get,

$$\Phi(z, \ldots, z) = \sum_{\substack{\lambda_0, \ldots, \lambda_n \\ =0}}^{L} p(\lambda_0, \ldots, \lambda_n) z^{\lambda_0} \left(\alpha_1^{\lambda_1} \ldots \alpha_n^{\lambda_n} \right)^z$$

$$= \sum_{\substack{\lambda_0, \ldots, \lambda_n \\ =0}}^{L} p(\lambda_0, \ldots, \lambda_n) z^{\lambda_0} \left(e^{\lambda_1 \log \alpha_1 + \cdots + \lambda_n \log \alpha_n} \right)^z$$

$$= \sum_{\alpha=0}^{RS-1} p_\alpha z^{\nu_\alpha} e^{\psi_\alpha z},$$

as desired. $\qquad \square$

Expressing the Coefficients p_α in Terms of Φ

Lemma 8.2.4 *Suppose there exists a coefficient $p(\lambda_0, \ldots, \lambda_n) \neq 0$ for some tuple $(\lambda_0, \ldots, \lambda_n) \in [0, L]^{n+1}$. Let r' be the integer given by*

$$r' = \lambda_0 + \lambda_1 S + \cdots + \lambda_n S^n.$$

Then there exist integers r, s with $0 \leq r \leq R - 1$ and $0 \leq s \leq S - 1$ such that

$$p(\lambda_1, \ldots, \lambda_n) = p_{r'} = \sum_{j=0}^{RS-1} c_j \left(\frac{d}{dz} \right)^j (\Phi(z, \ldots, z)) |_{z=0} = \sum_{j=0}^{RS-1} c_j f^{(j)}(0)$$

where c_j's are the coefficients of the augmentative polynomial $W(z)$ as given in (8.30).

Proof Note that $0 \leq r' \leq RS - 1$. Consider $\psi_{r'} = \lambda_1 \log \alpha_1 + \cdots + \lambda_n \log \alpha_n$ and $\nu_{r'} = \lambda_0$. If $r' \leq R - 1$, then we take $r = r'$. Let $r' > R - 1$. Since the distinct elements of $\mathcal{S}$ are listed as $\psi_0, \ldots, \psi_{R-1}$, by Lemma 8.2.1, there exists $r \leq R - 1$ such that $\psi_r = \psi_{r'}$. Take $s = \nu_{r'} = \lambda_0 \leq L \leq S - 1$. With these choices of integers (r, s) and $\psi_0, \ldots, \psi_{R-1}$ distinct complex numbers, there exists a polynomial $W(z) = \sum_{j=0}^{RS-1} c_j z^j$ as in Lemma 8.1.4, satisfying

(a) $|c_i| < (2\sigma/\rho)^{RS}$;

(b) $\dfrac{d^j}{dz^j} (W(z)) |_{z=\psi_i} = \begin{cases} 1 & \text{if } (i, j) = (r, s) \\ 0 & \text{if } (i, j) \neq (r, s) \end{cases}$

Here $\sigma = \max\{1, |\psi_i| : 0 \leq i \leq R - 1\}$ and $\rho = \min\{1, |\psi_i - \psi_j| : 0 \leq i < j \leq R - 1\}$. By the explicit values of (r, s), we rewrite (b) as follows.

$$\left(\frac{d}{dz} \right)^{\nu_j} (W(z)) |_{z=\psi_i} = \begin{cases} 1 & \text{if } \psi_i = \psi_{r'} \text{ and } \nu_j = \nu_{r'} \\ 0 & \text{if } \psi_i \neq \psi_{r'} \text{ or } \nu_j \neq \nu_{r'} \end{cases}$$

Suppose $\psi_i = \psi_j$ for some $0 \leq i \neq j \leq RS - 1$. Then $\nu_i \neq \nu_j$. Therefore, we can write

$$p_{r'} = \sum_{i=0}^{RS-1} p_i \left(\frac{d}{dz}\right)^{\nu_i} (W(z))|_{z=\psi_i}.$$

Now by Lemma 8.2.2,

$$
\begin{aligned}
p_{r'} &= \sum_{i=0}^{RS-1} p_i \left(\frac{d}{dz}\right)^{\nu_i} (W(z))|_{z=\psi_i} \\
&= \sum_{i=0}^{RS-1} p_i \sum_{j=0}^{RS-1} c_j \left(\frac{d}{dz}\right)^{j} (z^{\nu_i} e^{\psi_i z})|_{z=0} \\
&= \sum_{j=0}^{RS-1} c_j \left(\frac{d}{dz}\right)^{j} \left(\sum_{i=0}^{RS-1} p_i z^{\nu_i} e^{\psi_i z}\right)|_{z=0} \\
&= \sum_{j=0}^{RS-1} c_j \left(\frac{d}{dz}\right)^{j} (\Phi(z,\ldots,z))|_{z=0},
\end{aligned}
$$

by Lemma 8.2.3. This proves the lemma.   □

Finally...

Let r' be as given in Lemma 8.2.4. We will show that $|p_{r'}| < 1$. This will complete the proof of Baker's theorem. By Lemma 8.2.4, we have

$$|p_{r'}| \le \sum_{j=0}^{RS-1} |c_j||f_j(0)|,$$

where $RS = (L+1)^{n+1} \le (h^{2-\frac{1}{4n}} + 1)^{n+1} < h^{4n}$ and $|c_j| < (2\sigma/\rho)^{RS}$ where σ and ρ are defined as in Lemma 8.1.4. Firstly,

$$|\psi_i| \le |\lambda_1^{(i)}||\log\alpha_1| + \cdots + |\lambda_n^{(i)}||\log\alpha_n| \le L(|\log\alpha_1| + \cdots + |\log\alpha_n|) \le c_{8.36}L.$$

Hence

$$\sigma \le c_{8.36}L.$$

Now, we compute ρ as follows.

$$
\begin{aligned}
\rho &= \min_{0 \le i < j \le R-1} \{1, |\psi_i - \psi_j|\} \\
&= \min_{0 \le i < j \le R-1} \{1, |(\lambda_1^{(i)} - \lambda_1^{(j)})\log\alpha_1 + \cdots + (\lambda_n^{(i)} - \lambda_n^{(j)})\log\alpha_n|\}.
\end{aligned}
$$

Let $\rho \ne 1$. Then

$$\rho = |(\lambda_1^{(i)} - \lambda_1^{(j)}) \log \alpha_1 + \cdots + (\lambda_n^{(i)} - \lambda_n^{(j)}) \log \alpha_n|.$$

Putting $\kappa_k = \lambda_k^{(i)} - \lambda_k^{(j)}$ for $1 \le k \le n$ we get

$$\rho = |\kappa_1 \log \alpha_1 + \cdots + \kappa_n \log \alpha_n|, \tag{8.33}$$

for some integers $|\kappa_i| \le L$ and not all of them are zero since $i \ne j$. Therefore by Lemma 8.1.3, we get

$$\rho \ge c_{8.37}^{-L}. \tag{8.34}$$

This estimate also holds if $\rho = 1$ by taking $c_{8.37} > 1$.

Thus, for all integers j with $0 \le j < RS$, we have

$$|c_j| \le (2c_{8.36} L c_{8.37}^L)^{RS} \le L^{RS} c_{8.38}^{LRS}. \tag{8.35}$$

Thus, using also Lemma 8.1.8, we get

$$1 \le |p_{r'}| \le \sum_{j=0}^{RS-1} |c_j| |f^{(j)}(0)|$$
$$\le RSL^{RS} c_{8.38}^{LRS} e^{-h^{8n}}$$
$$\le h^{4n} h^{2h^{4n}} c_{8.38}^{h^{4n+2}} e^{-h^{8n}}$$
$$< 1,$$

as

$$4n \log h + 2h^{4n} \log h + h^{4n+2} \log c_{8.38} < c_{8.39} h^{4n+2} < h^{8n}$$

by taking $h > c_{8.39}^{1/(4n-2)}$. This proves Baker's theorem. $\qquad\square$

Exercise

(1) Let $x, y > 0$ be real numbers. Show that

$$x > \log(1 + x) > y(1 - (1 + x)^{-1/y}).$$

(2) Prove the Leibnitz's rule: If $f_j(x)$ has n derivatives for $1 \le j \le m$, show that

$$\frac{\partial^n}{\partial x^n} \prod_{j=1}^{m} f_j(x) = \sum \frac{i!}{i_1! \cdots i_m!} \frac{\partial^{i_1} f_1}{\partial x^{i_1}} \cdots \frac{\partial^{i_m} f_m}{\partial x^{i_m}}$$

with summation over all $i_1, \ldots, i_m \ge 0$ with $i_1 + \cdots + i_m = i$.

(3) For a real number t, let $t^{1/3}$ denote the real cubic root of t. Let $\alpha = (2 + \sqrt{5})^{1/3}$ and $\beta = (2 - \sqrt{5})^{1/3}$. Are α and β linearly independent over $\mathbb{Q}$? Are $1, \alpha, \beta$ linearly independent over $\mathbb{Q}$?

Notes

The ground work shows the importance of some factors:

(i) Choice of parameters.
(ii) Liouville-type argument to get a non-trivial lower bound for a non-zero algebraic integer since its norm is ≥ 1.
(iii) Construction of auxiliary polynomial by solving linear equations.

The above three factors have been encountered in earlier chapters as well.

(iv) The role of the auxiliary polynomial Φ in several variables:

(a) The factor $(\log \alpha_1)^{m_1} \cdots (\log \alpha_{n-1})^{m_{n-1}}$ can be pulled out nicely (see (8.8) and (8.9)) giving rise to linear equations with algebraic coefficients.

(b) More importantly, in the extrapolation argument, the partial derivatives can be taken in any direction instead of just along the diagonal (see (8.16)).

The quantitative results are more involved as the dependence on α_is and β_js have to be kept track. Since the results of Baker, several mathematicians like Ramachandra, Shorey, Waldschmidt and others have worked on improving the lower bounds. As mentioned in Chap. 7, the best known result in the general case till date is due to Matveev [4] while for linear forms in two and three logarithms the reader may see [5, 6].

References

1. A. Baker, Linear forms in the logarithms of algebraic numbers. Mathematika **13**, 204–216 (1966); II **14**, 102–107 (1967); III **14**, 220–228 (1967)
2. A. Baker, *Transcendental Number Theory, Tracts* (Cambridge, 1975)
3. N.I. Fel'dman, Y.V. Nesterenko, *Transcendental Numbers*. Number Theory IV, Encyclopaedia of Mathematical Sciences, vol. 44 (Springer, Berlin, 1991)
4. E.M. Matveev, An explicit lower bound for a homogeneous rational linear form in logarithms of algebraic numbers. Izv. Math. **62**, 81–136 (1998)
5. M. Laurent, M. Mignotte, Y. Nesterenko, Formes linéaires en deux logarithmes et déterminants d'interpolation. J. Number Theory **55**, 285–321 (1995)
6. C.D. Bennett, J. Blass, A.M.W. Glass, D.B. Meronk, R.P. Steiner, Linear forms in the logarithms of three positive rational integers. J. Theo. Nombr. Bordeaux **9**, 97–136 (1997)

Chapter 9
Subspace Theorem

All knowledge that the world has ever received comes from the mind; the infinite library of universe is in our own mind

—Ramakrishna

Subspace theorem is a multidimensional extension of Roth's theorem developed by Schmidt in 1980. He introduced several new ideas, especially from the geometry of numbers. An important ingredient was the properties of *successive minima*. Since the proofs of his results are beyond the scope of this book, we limit ourselves to stating two versions of his results and derive Roth's theorem. See Sect. 9.1. In Sect. 9.2, we present some classical approximation results derived from Dirichlet's multidimensional theorem. Section 9.3 deals with the application of subspace theorem to simultaneous approximation of algebraic numbers by rationals. Sections 9.1–9.3 are based on results from Schmidt [1, 2].

Several innovative applications of this result have been recently found. Starting from 1998, Bugeaud, Corvaja, Luca, Zannier and others have applied the subspace Theorem 9.1.2 successfully to some number theoretic problems like the growth of greatest prime factor of $(ab + 1)(ac + 1)(bc + 1)$ where $a > b > c$ are integers. In Sect. 9.4, we present one such result.

9.1 Statement of Subspace Theorem

We denote an n-tuple of variables by $\mathbf{X} = (X_1, \ldots, X_n)$. By $L = L(\mathbf{X})$ we mean a linear form in n variables $X_1, \ldots, X_n$. For any $\mathbf{x} = (x_1, \ldots, x_n) \in \mathbb{R}^n$, by $|\mathbf{x}|$ we mean $\max(1, |x_1|, \ldots, |x_n|)$.

© Springer Nature Singapore Pte Ltd. 2020
S. Natarajan and R. Thangadurai, *Pillars of Transcendental Number Theory*,
https://doi.org/10.1007/978-981-15-4155-1_9

Theorem 9.1.1 *Let $n \geq 2$ be an integer. Let $L_1, \ldots, L_n$ be n linearly independent linear forms in n variables with real or complex algebraic coefficients. Let $\epsilon > 0$. Then the set of solutions $\mathbf{x} = (x_1, \ldots, x_n) \in \mathbb{Z}^n$ to the inequality*

$$\prod_{i=1}^{n} |L_i(\mathbf{x})| < |\mathbf{x}|^{-\epsilon} \tag{9.1}$$

lies in finitely many proper rational subspaces of $\mathbb{Q}^n$.

By a rational subspace of $\mathbb{Q}^n$, we mean a subspace that can be defined by linear equations with rational coefficients. Let us derive Roth's theorem from the above result. Let

$$n = 2, L_1(\mathbf{X}) = \alpha X_2 - X_1, L_2(\mathbf{X}) = X_2.$$

Then for $\alpha \in \mathbb{A}$ and for any $\epsilon > 0$ the above theorem implies that all the points $\mathbf{x} = (x_1, x_2) \in \mathbb{Z}^2$ for which

$$|\alpha x_2 - x_1||x_2| < |\mathbf{x}|^{-\epsilon} \tag{9.2}$$

lie on a finite set of lines $x_1 = kx_2$, $k \in \mathbb{Q}$. Suppose if (p_0, q_0) and (tp_0, tq_0), $t \in \mathbb{Z}$, are two points lying on the line $x_1 = kx_2$, then

$$|tq_0\alpha - tp_0| < \max(|tq_0|, |tp_0|)^{-\epsilon}|tq_0|^{-1}$$

implies

$$|t|^{2+\epsilon} < |q_0|^{-1-\epsilon}|q_0\alpha - p_0|^{-1},$$

i.e., $|t|$ is bounded from above. Hence each such line can contain only finitely many points satisfying the inequality (9.2). So we conclude that there are only finitely many solutions which give Roth's theorem.

The subspace theorem states that the set of solutions of (9.1) lie in a finite union of proper subspaces of $\mathbb{Q}^n$. But one may ask if indeed (9.1) has only *finitely many* solutions. For instance, suppose there is a non-zero $\mathbf{x_0}$ such that $L_1(\mathbf{x_0}) = 0$. Then for any $\lambda\mathbf{x_0}$, $\lambda \in \mathbb{Z}$, inequality (9.1) is satisfied. Thus there are infinitely many solutions to (9.1). In the case $n = 2$, we have seen in the derivation of Roth's theorem that there are only finitely many solutions to (9.1). If $n \geq 3$, then (9.1) may very well having infinitely many solutions. For example, consider the following linear forms.

$$L_1 = x_1 + \sqrt{2}x_2 + \sqrt{3}x_3; L_2 = x_1 + \sqrt{2}x_2 - \sqrt{3}x_3; L_3 = x_1 - \sqrt{2}x_2 - \sqrt{3}x_3.$$

Take the subspace of $\mathbb{Q}^3$ defined by $x_3 = 0$. Then consider the infinitely many solutions to Pell's equation $x_1^2 - 2x_2^2 = 1$, say with $x_1 > 0$ and $x_2 < 0$ so that $x_1 + \sqrt{2}x_2$ is small. Then for any of these infinitely many solutions

$$|L_1 L_2 L_3| = |x_1 + \sqrt{2}x_2| \leq (\max(|x_1|, |x_2|)^{-\epsilon}$$

for many choices of ϵ. Now we state another form of subspace theorem which is a special case of a result from [3].

Theorem 9.1.2 *Let* $n \geq 2$ *be an integer. Let* S *be a finite set of primes. Let* $L_{1,\infty}, \ldots, L_{n,\infty}$ *be* n *linearly independent linear forms in* n *variables with real algebraic coefficients. For any prime number* ℓ *in* S *let* $L_{1,\ell}, \ldots, L_{n,\ell}$ *be* n *linearly independent linear forms with integer coefficients. Let* $\epsilon > 0$ *be given. Then the set of solutions* $\mathbf{x} = (x_1, \ldots, x_n) \in \mathbb{Z}^n$ *to the inequality*

$$\prod_{i=1}^{n} |L_{i,\infty}(\mathbf{x})| \cdot \prod_{\ell \in S} \prod_{i=1}^{n} |L_{i,\ell}(\mathbf{x})|_\ell \leq |\mathbf{x}|^{-\epsilon}$$

lies in finitely many proper subspaces of $\mathbb{Q}^n$.

9.2 Dirichlet's Multidimensional Approximation Results

Before going for an application of subspace theorem, we shall present some classical results due to Dirichlet which motivate the ensuing application. We recall Theorem 5.1.1 due to Dirichlet. In this section we denote by $c_{9,\ldots} = c_{9,\ldots}(\cdots)$ positive numbers depending on certain parameters which are mentioned within brackets, and these numbers can be effectively computable.

Theorem 9.2.1 *Let* α *and* Q *be real numbers with* $Q > 1$. *Then there exist integers* p, q *such that* $1 \leq q < Q$ *and* $|\alpha q - p| \leq 1/Q$.

It follows from this theorem that the inequality $\left|\alpha - \frac{p}{q}\right| < \frac{1}{q^2}$ has solutions in integers p, q and in fact, there exist infinitely many coprime integers p, q with this property if α is irrational. This is not true if α is rational. See Corollary 5.1.2 and the Remark following it.

An irrational number α is defined to be *badly approximable* if there is a constant $c_{9,1} = c_{9,1}(\alpha) > 0$ such that

$$\left|\alpha - \frac{p}{q}\right| > \frac{c_{9,1}}{q^2} \tag{9.3}$$

for every rational p/q.

In 1842, Dirichlet gave an extension of Theorem 9.2.1 to higher dimension.

Theorem 9.2.2 *Suppose that* $\alpha_{ij}, 1 \leq i \leq n, 1 \leq j \leq m$ *are* nm *real numbers and* $Q > 1$. *Then there exist integers* $q_1, \ldots, q_m, p_1, \ldots, p_n$ *with*

$$1 \leq \max(|q_1|, \ldots, |q_m|) < Q^{n/m},$$

such that

$$|\alpha_{i1}q_1 + \cdots + \alpha_{im}q_m - p_i| \le \frac{1}{Q} \ for \ 1 \le i \le n.$$

Proof We may assume that Q is an integer. Otherwise one may work with $Q' = [Q] + 1$. Consider the tuples

$$(\{\alpha_{11}x_1 + \cdots + \alpha_{1m}x_m\}, \ldots, \{\alpha_{n1}x_1 + \cdots + \alpha_{nm}x_m\})$$

where each x_j is an integer satisfying

$$0 \le x_j < Q^{n/m}, \quad 1 \le j \le m.$$

Here $\{x\}$ denotes the fractional part of x. There are at least Q^n such tuples, and each such tuple lies in the unit cube U in $\mathbb{R}^n$. Along with the tuple $(1, \ldots, 1)$ the unit cube U contains $Q^n + 1$ such tuples. Divide U into Q^n pairwise disjoint subcubes of side length $1/Q$. So among the $Q^n + 1$ tuples under consideration, at least two tuples will lie in the same cube. Let us say such two tuples are

$$(\alpha_{11}x_1 + \cdots + \alpha_{1m}x_m - y_1, \ldots, \alpha_{n1}x_1 + \cdots + \alpha_{nm}x_m - y_n),$$

$$(\alpha_{11}x_1' + \cdots + \alpha_{1m}x_m' - y_1', \ldots, \alpha_{n1}x_1' + \cdots + \alpha_{nm}x_m' - y_n')$$

with $(x_1, \ldots, x_m) \ne (x_1', \ldots, x_m')$, $y_i = [\alpha_{11}x_1 + \cdots + \alpha_{1m}x_m]$ and $y_i' = [\alpha_{11}x_1' + \cdots + \alpha_{1m}x_m']$, $1 \le i \le n$. Taking $q_i = x_i - x_i'$ for $1 \le i \le m$ and $p_j = y_j - y_j'$ for $1 \le j \le n$, we get the assertion of the theorem. $\square$

We derive some consequences. Taking $m = 1$ we immediately get the following result on *simultaneous approximation*.

Theorem 9.2.3 *Suppose that $\alpha_1, \ldots, \alpha_n$ are n real numbers and that $Q > 1$ is an integer. Then there exist integers $q, p_1, \ldots, p_n$ with*

$$1 \le q < Q^n \ and \ |q\alpha_i - p_i| \le \frac{1}{Q} \ for \ 1 \le i \le n.$$

As seen in the derivation of Corollary 5.1.2, we can deduce from the above theorem the following result.

Corollary 9.2.4 *Suppose that at least one of $\alpha_1, \ldots, \alpha_n$ is irrational. Then there are infinitely many n-tuples $(p_1/q, \ldots, p_n/q)$ with*

$$\left| \alpha_i - \frac{p_i}{q} \right| < \frac{1}{q^{1+1/n}}, 1 \le i \le n.$$

Taking $n = 1$ in Theorem 9.2.2, we get the next result which deals with a single linear form in $\alpha_1, \ldots, \alpha_n$.

Theorem 9.2.5 *Suppose that $\alpha_1, \ldots, \alpha_n$ are n real numbers and that $Q > 1$ is an integer. Then there exist integers $q_1, \ldots, q_n, p$ with*

$$1 \leq \max(|q_1|, \ldots, |q_n|) < Q^{1/n} \text{ and } |\alpha_1 q_1 + \cdots + \alpha_n q_n - p| \leq \frac{1}{Q}.$$

As a consequence of the above theorem we get

Corollary 9.2.6 *Suppose $1, \alpha_1, \ldots, \alpha_n$ are linearly independent over the rationals. Then there are infinitely many coprime $(n+1)$-tuples $(q_1, \ldots, q_n, p)$ with*

$$q = \max(|q_1|, \ldots, |q_n|) > 0 \text{ and } |\alpha_1 q_1 + \cdots + \alpha_n q_n - p| < \frac{1}{q^n}.$$

Let us use some notation for further discussion. Let $\alpha_{ij}, 1 \leq i \leq n, 1 \leq j \leq m$ be nm real algebraic numbers. For any tuple $\mathbf{x} = (x_1, \ldots, x_m)$, with x_i real algebraic, define $\lceil \mathbf{x} \rceil = \max(\lceil x_1 \rceil, \ldots, \lceil x_m \rceil)$ and

$$L_i(\mathbf{x}) = \alpha_{i1} x_1 + \cdots + \alpha_{im} x_m \text{ for } 1 \leq i \leq n.$$

Further let
$$\mathfrak{L}(\mathbf{x}) = (L_1(\mathbf{x}), \ldots, L_n(\mathbf{x})).$$

With this notation, Theorem 9.2.2 implies that there exist integral tuples $\mathbf{q} = (q_1, \ldots, q_m)$ and $\mathbf{p} = (p_1, \ldots, p_n)$ such that

$$\left\lceil \mathfrak{L}(\mathbf{q}) - \mathbf{p} \right\rceil^n \lceil \mathbf{q} \rceil^m < 1.$$

We say that $(L_1, \ldots, L_n)$ is a *badly approximable system of linear forms* if there is a constant $c_{9.2} = c_{9.2}(L_1, \ldots, L_n) > 0$ such that

$$\lceil \mathbf{x} \rceil^m \left\lceil \mathfrak{L}(\mathbf{x}) - \mathbf{y}^n \right\rceil > c_{9.2} \tag{9.4}$$

for every integer tuples $\mathbf{x}$ and $\mathbf{y}$ with $\mathbf{x} \neq \mathbf{0}$.

Suppose $m = n = 1$. Then (9.4) gives that the linear form $L(x) = \alpha_{11} x - y$ is badly approximable if

$$\left| \alpha_{11} - \frac{y}{x} \right| > \frac{c_{9.2}}{x^2}.$$

Thus α_{11} is badly approximable in the sense of (9.3).

Suppose $n = 1$. Then a single linear form $L(\mathbf{x})$ is badly approximable if

$$|\alpha_1 q_1 + \cdots + \alpha_m q_m - p| > \frac{c_{9.3}}{q^m}$$

for every integer point $(q_1, \ldots, q_m, p)$ with $q = \max(|q_1|, \ldots, |q_m|) > 0$ and $c_{9.3} = c_{9.3}(L)$.

Suppose $m = 1$. Then we have $L_1(x) = \alpha_1 x, \ldots, L_n(x) = \alpha_n x$ and

$$\max(|\alpha_1 q - p_1|, \ldots, |\alpha_n q - p_n|) > \frac{c_{9.4}}{q^{1/n}}$$

for integers $q > 0$, $p_1, \ldots, p_n$ with $c_{9.4} = c_{9.4}(L_1, \ldots, L_n)$. In other words,

$$\max\left(\left|\alpha_1 - \frac{p_1}{q}\right|, \ldots, \left|\alpha_n - \frac{p_n}{q}\right|\right) > \frac{c_{9.4}}{q^{1+1/n}}.$$

In this case, we say the tuple $(\alpha_1, \ldots, \alpha_n)$ is a *badly approximable* tuple.

We now show the following result.

Theorem 9.2.7 *Suppose $1, \alpha_1, \ldots, \alpha_m$ is a basis of a real algebraic number field of degree $m + 1$. Then the linear form $\alpha_1 x_1 + \cdots + \alpha_m x_m$ is badly approximable.*

Proof We denote by $c_{9.5}, c_{9.6}, \ldots$ positive numbers which depend only on $\alpha_1, \ldots, \alpha_m$. We may restrict to integers $q_1, \ldots, q_m, p$ with $|\alpha_1 q_1 + \cdots + \alpha_m q_m - p| < 1$ since otherwise the assertion is obviously true. Then

$$|p| \leq c_{9.5} q$$

where $q = \max(|q_1|, \ldots, |q_m|)$. Also each conjugate

$$|\alpha_1^{(i)} q_1 + \cdots + \alpha_m^{(i)} q_m - p| \leq c_{9.6} q.$$

Hence the norm

$$|\mathcal{N}(\alpha_1 q_1 + \cdots + \alpha_m q_m - p)| \leq c_{9.7} q^m |\alpha_1 q_1 + \cdots + \alpha_m q_m - p|. \qquad (9.5)$$

Let d be the denominator of $\alpha_1, \ldots, \alpha_m$. Then

$$|\mathcal{N}(d\alpha_1 q_1 + \cdots + d\alpha_m q_m - dp)| \geq 1$$

which gives $|\mathcal{N}(\alpha_1 q_1 + \cdots + \alpha_m q_m - p)| \geq d^{-m-1} = c_{9.8}$ which together with (9.5) yields the result. $\qquad\square$

By taking $q_{v+1} = \cdots = q_m = 0$, we may state the above theorem in a slightly more general way as follows.

Theorem 9.2.8 *Suppose $1, \alpha_1, \ldots, \alpha_v$ are linearly independent over $\mathbb{Q}$ and they generate an algebraic number field of degree v. Then there exists a number $c_{9.9} = c_{9.9}(\alpha_1, \ldots, \alpha_v)$ such that*

$$|\alpha_1 q_1 + \cdots + \alpha_v q_v - p| > c_{9.9} q^{-v+1}$$

for any integers $q_1, \ldots, q_v, p$ with $q = \max(|q_1|, \ldots, |q_v|) > 0$.

When $v = 1$, we obtain Liouville's theorem. When $v = \nu - 1$, the exponent is best possible by Theorem 9.2.7. So far, note that the results are *effective* in the sense that the numbers $c_{9,...}$ are computable. When $v < \nu - 1$, the exponent can be improved as shown in Theorem 9.3.1 below.

9.3 Applications of Subspace Theorems to Diophantine Approximation

We state and prove *Roth-type* results as application of subspace Theorem 9.1.1. Wherever subspace theorem is applied, the reader should note that the results are *ineffective*. Henceforth, we shall denote by $\gamma_{9,...} = \gamma_{9,...}(...)$ numbers which are *ineffective*, i.e. which cannot be effectively computable. On the other hand, $c_{9,10}, \ldots$ are effectively computable numbers depending only on $\alpha_1, \ldots, \alpha_n$.

Theorem 9.3.1 *Suppose $\alpha_1, \ldots, \alpha_u$ are real algebraic numbers such that $1, \alpha_1, \ldots, \alpha_u$ are linearly independent over $\mathbb{Q}$. Let $\epsilon > 0$. Then there are only finitely many positive integers q with*

$$q^{1+\epsilon}\|\alpha_1 q\| \cdots \|\alpha_u q\| < 1 \tag{9.6}$$

where $\|x\|$ denotes the distance of x to the nearest integer.

Proof Let q satisfy (9.6). Choose $p_1, \ldots, p_u$ with $\|\alpha_i q\| = |\alpha_i q - p_i|$ for $1 \leq i \leq u$. Note that $|p_i| \leq |\alpha_i| q + 1/2, 1 \leq i \leq u$. Put $n = u + 1$ and take

$$\mathbf{x_0} = (x_1, \ldots, x_n) = (p_1, \ldots, p_u, q).$$

Then $q \leq |\mathbf{x_0}| \leq c_{9,10} q$. Let us introduce linear forms

$$L_i(\mathbf{x}) = \alpha_i X_n - X_i, 1 \leq i \leq u; \quad L_n(\mathbf{x}) = X_n.$$

Then (9.6) implies that

$$|L_1(\mathbf{x_0}) \cdots L_n(\mathbf{x_0})| < q^{-\epsilon} \tag{9.7}$$
$$\leq c_{9,10}^{\epsilon} |\mathbf{x_0}|^{-\epsilon}$$
$$\leq |\mathbf{x_0}|^{-\epsilon/2}$$

whenever $q > c_{9,10}^2$. Consider

$$|L_1(\mathbf{x}) \cdots L_n(\mathbf{x})| < |\mathbf{x}|^{-\epsilon/2} \tag{9.8}$$

By Theorem 9.1.1, the solutions to (9.8) lie in finitely many proper rational subspaces, and by (9.7), $\mathbf{x_0}$ is a solution of (9.8). Let T be a subspace containing $\mathbf{x_0}$. Let it be defined by

$$c_1 x_1 + \cdots + c_u x_u + c_n x_n = 0.$$

Then for $\mathbf{x} = (p_1, \ldots, p_u, q)$ in T we have

$$c_1(\alpha_1 q - p_1) + \cdots + c_u(\alpha_u q - p_u) = (c_1 \alpha_1 + \cdots + c_u \alpha_u + c_n) q.$$

Hence

$$|c_1| \| \alpha_1 q \| + \cdots + |c_u| \| \alpha_u q \| \geq \gamma_{9.1} q \qquad (9.9)$$

where $\gamma_{9.1} = |c_1 \alpha_1 + \cdots + c_u \alpha_u + c_n|$. Note that $\gamma_{9.1} > 0$ since $\alpha_1, \ldots, \alpha_u$ are linearly independent. Also $\gamma_{9.1}$ is *ineffective* since we only know the existence of $c_1, \ldots, c_u, c_n$. From (9.9), we therefore get that

$$q \leq \frac{|c_1| + \cdots + |c_u|}{2\gamma_{9.1}}$$

showing that q is bounded.   □

Suppose

$$\left| \alpha_i - \frac{p_i}{q} \right| < q^{-1-(1/u)-\epsilon}, \ 1 \leq i \leq u. \qquad (9.10)$$

This implies that

$$\| \alpha_i q \| < q^{-1/u - \epsilon}, \ 1 \leq i \leq u.$$

Hence (9.6) is satisfied. So (9.10) has only finitely many solutions. Thus we get the following result.

Corollary 9.3.2 *Suppose $\alpha_1, \ldots, \alpha_u$ are real algebraic numbers such that $1, \alpha_1, \ldots, \alpha_u$ are linearly independent over $\mathbb{Q}$. Let $\epsilon > 0$. Then there are only finitely many rational u-tuples $\left(\frac{p_1}{q}, \ldots, \frac{p_u}{q} \right)$ satisfying (9.10).*

Theorem 9.3.3 *Suppose $\alpha_1, \ldots, \alpha_u$ are real algebraic numbers such that $1, \alpha_1, \ldots, \alpha_u$ are linearly independent over $\mathbb{Q}$. Let $\epsilon > 0$. Then there are only finitely many v-tuples of non-zero integers $(q_1, \ldots, q_v)$ with*

$$|q_1 \cdots q_v|^{1+\epsilon} \| \alpha_1 q_1 + \cdots + \alpha_v q_v \| < 1. \qquad (9.11)$$

Remark

We know that

$$\| \alpha_1 q_1 + \cdots + \alpha_v q_v \| < 1.$$

Taking $q = \max(|q_1|, \ldots, |q_v|)$, we get, under the hypothesis of Theorem 9.3.3, that there exists an integer p such that

$$\|\alpha_1 q_1 + \cdots + \alpha_v q_v\| = |\alpha_1 q_1 + \cdots + \alpha_v q_v - p| < q^{-v-\epsilon}.$$

The exponent is *best possible* by Theorem 9.2.8. Now we prove Theorem 9.3.3.

Proof Suppose $q_1, \ldots, q_v$ are given integers satisfying (9.11). Choose p with

$$\|\alpha_1 q_1 + \cdots + \alpha_v q_v\| = |\alpha_1 q_1 + \cdots + \alpha_v q_v - p|.$$

Then

$$|\alpha_1 q_1 + \cdots + \alpha_v q_v - p| \leq 1/2$$

giving

$$|p| \leq c_{9.11} q$$

where $c_{9.11} = \max(1, |\alpha_1|, \ldots, |\alpha_v|)$. Put $n = v + 1$ and write

$$\mathbf{x_0} = (q_1, \ldots, q_v, p).$$

Then $q \leq |\mathbf{x_0}| \leq c_{9.11} q$. Let us introduce linear forms

$$L_i(\mathbf{x}) = X_i, \ 1 \leq i \leq v; \ L_n(\mathbf{x}) = \alpha_1 X_1 + \cdots + \alpha_v X_v - X_n.$$

Then by Theorem 9.1.1, the solutions lie in a finite number of rational subspaces. Let a typical such subspace be given by $c_1 x_1 + \cdots + c_v x_v + c_n x_n = 0$ in which $\mathbf{x_0}$ lies.

Suppose $c_v \neq 0$. Then we get

$$
\begin{aligned}
c_v \|\alpha_1 q_1 + \cdots + \alpha_v q_v\| &= c_v |\alpha_1 q_1 + \cdots + \alpha_v q_v - p| \\
&= |(c_v \alpha_1 - c_1 \alpha_v) q_1 + \cdots + (c_v \alpha_{v-1} - c_{v-1} \alpha_v) q_{v-1} - (c_v + c_n \alpha_v) p| \\
&= |c_v + c_n \alpha_v| |\alpha_1' q_1 + \cdots + \alpha_{v-1}' q_{v-1} - p|
\end{aligned}
$$

where $\alpha_i' = (c_v \alpha_i - c_i \alpha_v)/(c_v + c_n \alpha_v)$, $1 \leq i \leq v$. We prove the theorem by induction on v. Suppose that $v = 1$. Then $|\alpha_1 q - p| < 1/2$ implies that $|\alpha_1 |q - |p|| < 1/2$ giving

$$|p| > c_{9.12} q \text{ for } q > c_{9.13}.$$

Further, we get $c_1 \|\alpha_1 q\| = |p(c_1 + c_2 \alpha_1)| \geq \gamma_{9.2} |p| \geq 1$ if $q > 1/(\gamma_{9.2} c_{9.12})$. For $v > 1$ we obtain

$$c_v \|\alpha_1 q_1 + \cdots + \alpha_v q_v\| \geq \gamma_{9.3} \|\alpha_1' q_1 + \cdots + \alpha_{v-1}' q_{v-1}\|.$$

Hence by (9.11), we get

$$|q_1 \cdots q_{v-1}|^{1+\epsilon/2} \|\alpha_1' q_1 + \cdots + \alpha_{v-1}' q_{v-1}\| < 1.$$

Since $1, \alpha_1', \ldots, \alpha_{v-1}'$ are linearly independent over $\mathbb{Q}$ this inequality has only finitely many solutions, by induction.

Similar argument holds if $c_j \neq 0, 1 \leq j \leq v$ or if $c_1 = \cdots = c_v = 0$ but $c_n \neq 0$. $\square$

9.4 A Different Application

Here we present a result of Bugeaud et al. [4].

Theorem 9.4.1 *Let a, b be multiplicatively independent integers ≥ 2 and let $\epsilon > 0$. Then there exists a positive number n_0 such that*

$$\gcd(a^n - 1, b^n - 1) < \exp(\epsilon n)$$

whenever $n \geq n_0$.

Remarks

(1) The theorem implies that $a^n - 1$ and $b^n - 1$ cannot have a common factor of significant size.
(2) Let $g = \gcd(a^n - 1, b^n - 1)$. Write

$$\frac{b^n - 1}{a^n - 1} = \frac{c}{d}$$

with $\gcd(c, d) = 1$. Suppose $d \geq a^{(1-\epsilon)n}$, then Theorem 9.4.1 follows since

$$g = \frac{a^n - 1}{d} \leq a^{\epsilon n}.$$

(3) The number n_0 cannot be computed due to the ineffective nature of subspace theorem.

Proof As seen in Remark (2), it is enough to show that

$$d \geq a^{(1-\epsilon)n}.$$

Setting
For integers $j \geq 1$, let

$$z_j(n) = \frac{b^{jn} - 1}{a^n - 1} = \frac{c_{j,n}}{d_n}$$

with $\gcd(c_{j,n}, d_n) = 1$. Since

$$z_j(n) = \frac{b^{jn} - 1}{b^n - 1} z_1(n)$$

and

$$\frac{b^{jn} - 1}{b^n - 1} \in \mathbb{Z}$$

we see that $d_n = d_1 = d$. We shall assume that

$$d_n \le a^{(1-\epsilon)n}$$

for infinitely many values of n and arrive at a contradiction. We fix two positive integers k and h satisfying

$$(i) \ k > 2/\epsilon \quad \text{and} \quad (ii) \ a^h > 2a^k b^{k^2}. \tag{9.12}$$

Approximation for $z_j(n)$

We have

$$\frac{1}{a^n - 1} = \sum_{r=1}^{\infty} \frac{1}{a^{rn}} = \sum_{r=1}^{h} \frac{1}{a^{rn}} + O\left(a^{-(h+1)n}\right)$$

giving

$$\frac{b^{jn} - 1}{a^n - 1} = (b^{jn} - 1)\left(\sum_{r=1}^{h} \frac{1}{a^{rn}} + O\left(a^{-(h+1)n}\right)\right).$$

Thus

$$\left| z_j(n) + \sum_{r=1}^{h} \frac{1}{a^{rn}} - \sum_{r=1}^{h} \left(\frac{b^j}{a^r}\right)^n \right| = O\left(b^{jn} a^{-(h+1)n}\right). \tag{9.13}$$

Linear Forms for Applying Theorem 9.1.2

We take

$$S = \{\text{primes dividing } ab\}.$$

Let

$$N = hk + h + k.$$

For any $\mathbf{x} \in \mathbb{R}^N$, we write

$$\mathbf{x} = (x_1, \ldots, x_N) = (z_1, \ldots, z_k, y_{01}, \ldots, y_{0h}, \ldots, y_{k1}, \ldots, y_{kh}).$$

We take the following N linearly independent linear forms:

$$L_{i,\infty}(\mathbf{x}) = z_i + y_{01} + \cdots + y_{0h} - y_{i1} - \cdots - y_{ih} \text{ for } 1 \le i \le k;$$

for $(i, v) \notin \{(1, \infty), \ldots, (k, \infty)\}$, take

$$L_{i,v}(\mathbf{x}) = x_i.$$

The Smallness of the Linear Forms at $\mathbf{x_0}$
We consider the point $\mathbf{x_0}$ where

$$\begin{cases} z_i = d_n a^{hn} z_i(n) \text{ for } 1 \le i \le k, \\ y_{it} = d_n a^{hn} (b^i a^{-t})^n \text{ for } 0 \le i \le k; 1 \le t \le h. \end{cases}$$

Let $A = a^{h+1} b^k$. Note that

$$|z_j| = |d_n a^{hn} z_j(n)| \le d_n a^{hn} b^{jn} \le A^n \text{ for } 1 \le j \le k;$$

$$|y_{it}| \le d_n (a^h b^k)^n \le A^n \text{ for } 0 \le i \le k; 1 \le t \le h.$$

Hence

$$|\mathbf{x_0}| \le A^n.$$

Further by the product formula we have

$$\prod_{v \in S \cup \{\infty\}} |m|_v = 1$$

for any rational m composed of only the primes dividing a and b. For $i > k$, consider

$$\prod_{v \in S \cup \{\infty\}} |L_{i,v}(\mathbf{x_0})|_v = \prod_{v \in S \cup \{\infty\}} |d_n a^{hn} (b^i a^{-t})^n|_v \qquad (9.14)$$

$$= \prod_{v \in S \cup \{\infty\}} |d_n|_v$$

$$\le |d_n|_\infty = d_n.$$

For $1 \le i \le k$, $x_i = d_n a^{hn} z_i(n)$ which gives

$$\prod_{p|ab} |x_i|_p \le a^{-hn}. \qquad (9.15)$$

Further by (9.13), for $1 \le i \le k$, we have

$$|L_{i,\infty}(\mathbf{x_0})| = |z_i + y_{01} + \cdots + y_{0h} - y_{i1} - \cdots - y_{ih}|$$
$$= |d_n a^{hn}||z_i(n) + a^{-n} + \cdots + a^{-hn} - (b^i a^{-1})^n - \cdots - (b^i a^{-h})^n|$$
$$= O(b^{in} a^{-hn-n} d_n a^{hn})$$
$$= O(b^{in} a^{-n} d_n).$$

Combining this with (9.14) and (9.15) we get

$$\prod_{v \in S \cup \{\infty\}} \prod_{i=1}^{N} |L_{i,v}(\mathbf{x_0})|_v \leq d_n^{N-k} \prod_{v \in S \cup \{\infty\}} \prod_{i=1}^{k} |L_{i,v}(\mathbf{x_0})|_v$$

$$= d_n^{N-k} \prod_{i=1}^{k} |L_{i,\infty}(\mathbf{x_0})| \prod_{p|ab} \prod_{i=1}^{k} |x_i|_p$$

$$= O(d_n^N b^{k^2 n} a^{-hkn})$$

$$= O(a^{(1-\epsilon)nN} b^{k^2 n} a^{-hkn})$$

$$= O((b^{k^2} a^{h+k} a^{-\epsilon N})^n)$$

$$= O(2^{-n})$$

by the choice of k and h in (9.12). Let us take $\delta < (\log 2)/(\log A)$. Thus

$$\prod_{v \in S \cup \{\infty\}} \prod_{i=1}^{N} |L_{i,v}(\mathbf{x_0})|_v = O(A^{-\delta n}) = O(|\mathbf{x_0}|)^{-\delta}.$$

This is true for infinitely many n.

Application of Theorem 9.1.2
By subspace Theorem 9.1.2, $\mathbf{x_0}$ must lie on finitely many proper subspaces of $\mathbb{Q}^N$. Hence there is a proper subspace having infinitely many such $\mathbf{x_0}$. So for infinitely many n, such $\mathbf{x_0}$ must lie in the hyper plane say,

$$\zeta_1 Z_1 + \cdots + \zeta_k Z_k + \sum_{i,j} \alpha_{ij} Y_{ij} = 0$$

i.e.

$$\zeta_1 \frac{b^n - 1}{a^n - 1} + \cdots + \zeta_k \frac{b^{kn} - 1}{a^n - 1} + \sum_{i,j} \alpha_{ij} \left(\frac{b^j}{a^i}\right)^n = 0 \tag{9.16}$$

valid for infinitely many n. Let $\mathcal{A}$ denote the set of these infinitely many values of n.

Algebraic Independence of the Functions a^z and b^z
Suppose a^z and b^z are algebraically dependent. Then there exists a relation

$$\sum c_{ij}(a^i b^j)^z = 0 \tag{9.17}$$

with c_{ij} not all zero. Since a and b are multiplicatively independent, the numbers $a^i b^j$ are distinct. Let $a^{i_0} b^{j_0}$ be of the largest value among them with $c_{i_0 j_0} \neq 0$. Now let z run through the integer values in $\mathcal{A}$. Since $\mathcal{A}$ is an infinite set, we find that the left-hand side of (9.17) tends to $c_{i_0 j_0} \neq 0$, a contradiction.

Final Contradiction

Since a^z and b^z are algebraically independent, (9.16) gives rise to an identity

$$\zeta_1 \frac{Y-1}{X-1} + \cdots + \zeta_k \frac{Y^k-1}{X-1} + \sum_{i,j} \alpha_{ij} \left(\frac{Y^j}{X^i}\right)^n = 0$$

in $\mathbb{Q}(X, Y)$. This may be rewritten as

$$\frac{f(Y)}{X-1} + \frac{g(X,Y)}{X^h} = 0$$

where

$$f(Y) = \zeta_1(Y-1) + \cdots + \zeta_k(Y^k - 1)$$

and $g(X, Y) = \sum_{i,j} \alpha_{ij} X^{h-i} Y^j$. Thus $X - 1$ divides $f(Y)$ in $\mathbb{Q}[X, Y]$ which means $f(Y) = 0$ and hence $g(X, Y) = 0$. This implies that $\zeta_1 = \cdots = \zeta_k = 0$ and $\alpha_{ij} = 0$ for all i, j giving the final contradiction. $\qquad\square$

Exercise

1. Prove that if a and b are multiplicatively independent, a^z and b^z are algebraically independent functions using Lemma 1.1.2.
2. Let S be the set containing all the integers composed of given primes. Assume that a relation of the type

$$\sum_{ij} \gamma_{i,j} s_1^i s_2^j = 0, \gamma_{ij} \in \mathbb{Q} \tag{9.18}$$

does not hold for infinitely many pairs of multiplicatively independent integers $(s_1, s_2) \in S^2$ with $|s_2|^{\epsilon/2} \le |s_1| < |s_2|$. Show that for any $\epsilon > 0$,

$$\gcd(s_1 - 1, s_2 - 1) < \max(|s_1|, |s_2|)^\epsilon \tag{9.19}$$

for all pairs $(s_1, s_2) \in S$ with $\min(|s_1|, |s_2|) > 1$ and $\max(|s_1|, |s_2|)$ exceeding some number depending on ϵ and $\log|s_2|/\log|s_1| \notin \mathbb{Q}$. (Hint: Follow the proof of Theorem 9.4.1 and prove (9.19).)
3. Let $0 < \delta < 1$. For $\mathbf{x} = (x_1, x_2, x_3) \in \mathbb{Z}^3$, consider

$$0 < |(x_1 + \sqrt{2}x_2 + \sqrt{3}x_3)(x_1 - \sqrt{2}x_2 + \sqrt{3}x_3)(x_1 - \sqrt{2}x_2 - \sqrt{3}x_3)| \le |\mathbf{x}|^{-\delta}. \tag{9.20}$$

   (a) Prove that (9.20) has infinitely many solutions in the spaces $x_1 = 0$ and $x_2 = 0$.
   (b) Prove that (9.20) has only finitely many solutions with $x_1 x_2 x_3 \ne 0$.

Notes

Schmidt himself gave an upper bound for the number of subspaces in Theorem 9.1.1. Such results are known as quantitative subspace theorems. He applied his result to estimate the number of solutions of norm form equations. Schlickewei generalised the theorem to an arbitrary absolute value—p-adic subspace theorem and also to algebraic integer solutions. These results have been applied to the study of various Diophantine equations. See [3, 5, 6]. Important contributions towards this area were also made by Evertse, Győry, Thunder and others.

Hernández and Luca [7] generalised Theorem 9.4.1 to S-units. They showed that the assertion (9.19) in Exercise 2 above holds even if the condition (9.18) does not hold. As an application of their result, they could resolve a conjecture of Győry et al. [8] that for positive integers $a > b > c$ we have

$$\lim_{a \to \infty} P((ab + 1)(ac + 1)(bc + 1)) = \infty.$$

Here $P(n)$ denotes the greatest prime factor of an integer $n > 1$.

As mentioned earlier, Corvaja and Zannier applied successfully, various versions of subspace theorem to many number theoretic problems. Here we mention two of their interesting results. Let $\theta > 1$. The distribution of the sequence $\{\theta^n\}$ where $\{x\}$ denotes the fractional part of x is one of the intriguing problems in number theory. It is well known that $\{\theta^n\}$ is uniformly distributed in $[0, 1]$ for almost all real $\theta > 1$. On the other hand, it is *not* known whether $\{(3/2)^n\}$ is dense in $[0, 1]$. Mahler [9] showed that for any rational θ and $0 < \ell < 1$,

$$\{\theta^n\} \leq \ell^n \tag{9.21}$$

for all but a finite set of integers n depending on θ and ℓ. His result depends on Theorem 6.5.2. Similar result does not hold if θ were a suitable algebraic number, for instance, if θ is a Pisot number. Mahler asked for what algebraic numbers similar result holds. In [10], Corvaja and Zannier answered Mahler's question by showing a best possible result that if (9.21) holds for infinitely many n, then θ^d is a Pisot number for some positive integer d.

As another application, they showed that the length of the period of the continued fraction expansion of $\alpha^n \to \infty$ as $n \to \infty$. Here α is a real quadratic irrational which is not a square root of a rational number and not a unit in $\mathcal{O}_{\mathbb{Q}(\alpha)}$. This solves a question of Mendés France.

We strongly urge the reader to browse through the papers of Corvaja, Zannier and their co-authors for a variety of other results.

References

1. W.M. Schmidt, *Diophantine Approximation*. LNM , vol. 785 (Springer, Berlin, 1980)
2. W.M. Schmidt, *Diophantine Approximations and Diophantine Equations*. LNM, vol. 1467 (Springer, Berlin, 1991)
3. H.P. Schlickewei, The 𝔓-adic Thue-Siegel-Roth-Schmidt theorem. Arch. Math. (Basel) **29**, 267–270 (1977)

4. Y. Bugeaud, P. Corvaja, U. Zannier, An upper bound for the G.C.D of $a^n - 1$ and $b^n - 1$. Math. Z. **243**, 79–84 (2003)

5. H.P. Schlickewei, Die p-adische Verallgemeinerung des Satzes von Thue-Siegel-Roth-Schmidt. J. Reine Angew. Math **288**, 86–105 (1976)

6. H.P. Schlickewei, Linearformen mit algebraischen koeffizienten. Manuscripta Math. **18**, 147–185 (1976)

7. S. Hernández, F. Luca, On the largest prime factor of $(ab + 1)(ac + 1)(bc + 1)$. Bol. Soc. Mat. Mexicana (3) **9**, 235–244 (2003)

8. K. Győry, A. Sárközy, C.L. Stewart, On the number of prime factors of integers of the form $ab + 1$. Acta Arith. **74**(4), 365–385 (1996)

9. K. Mahler, On the fractional parts of the powers of a rational number II. Mathematika **4**, 122–124 (1957)

10. P. Corvaja, U. Zannier, On the rational approximation to the powers of an algebraic number: Solution of two problems of Mahler and Mendés-France. Acta Math. **193**, 175–191 (2004)

Appendix A
Introductory Quotes

Later on, when they had all said "Good-bye" and "Thank You" to Christopher Robin, Pooh and Piglet walked home thoughtfully together in the golden evening and for a long time they were silent. 'When you wake up in the morning, Pooh, 'said Piglet at last, "What's the first thing you say to yourself?" "What's for breakfast?" said Pooh. "What do you say, Piglet?" " I say, I wonder what's going to happen exciting to-day?" said Piglet. Pooh nodded thoughtfully. "It's the same thing," he said.

-Winnie the Pooh

© Springer Nature Singapore Pte Ltd. 2020
S. Natarajan and R. Thangadurai, *Pillars of Transcendental Number Theory*,
https://doi.org/10.1007/978-981-15-4155-1

Index

© Springer Nature Singapore Pte Ltd. 2020
S. Natarajan and R. Thangadurai, *Pillars of Transcendental Number Theory*,
https://doi.org/10.1007/978-981-15-4155-1

Printed in the United States
by Baker & Taylor Publisher Services